PNSO FIELD GUIDE TO THE ANCIENT WORLD

AGE
OF
DINOSAURS

AGE
OF
DINOSAURS

ZHAO Chuang

YANG Yang

BROWN BOOKS
PUBLISHING GROUP

PNSO Field Guide to the Ancient World: The Age of Dinosaurs

by: Yang Yang and Zhao Chuang

Brown Books Publishing Group
Dallas / New York
www.BrownBooks.com
(972) 381-0009

A New Era in Publishing®

Publisher's Cataloging-In-Publication Data

Names: Yang, Yang, 1982- author. | Zhao, Chuang, 1985- illustrator. | Chen, Mo (Translator), translator. | PNSO (Organization), production company.
Title: Age of dinosaurs / ZHAO Chuang [illustrator], YANG Yang [author] ; [translator, CHEN Mo].
Description: Dallas ; New York : Brown Books Publishing Group, [2021] | Series: PNSO field guide to the ancient world ; [1] | Translated from the Chinese, published in 2015. | Include bibliographical references and index.
Identifiers: ISBN 9781612545288
Subjects: LCSH: Dinosaurs--History. | Paleontology--Triassic. | LCGFT: Illustrated works.
Classification: LCC QE861.4 .Y36 2021 | DDC 567.9--dc23

ISBN 978-1-61254-528-8
LCCN 2021908154

Printed in China
10 9 8 7 6 5 4 3 2 1

For more information or to contact the author, please go to www.BrownBooks.com.

Dedicated to

Every living being that enriched the Earth

Contents

9 / The Lonesome Triassic Period

27 / The Bustling Jurassic Period

81 / The Desolate Cretaceous Period

FOREWORD

Dr. Mark A. Norell's Introduction to the Works by ZHAO Chuang and YANG Yang

Mark A. Norell is a renowed international paleontologist, Chair and Macaulay Curator for the Division of Paleontology of the AMNH, and science consultant for PNSO.

I am a paleontologist at one of the world's great museums. I get to spend my days surrounded by dinosaur bones. Whether it is in Mongolia excavating, in China studying, in New York analyzing data, or anywhere on the planet writing, teaching, or lecturing, dinosaurs are not only my interest, but my livelihood.

Most scientists, even the most brilliant ones, work in very closed societies. A system which, no matter how hard they try, is still unapproachable to average people. Maybe it's due to the complexities of mathematics, difficulties in understanding molecular biochemistry, or reconciling complex theory with actual data. No matter what, this behavior fosters boredom and disengagement. Personality comes in as well, and most scientists lack the communication skills necessary to make their efforts interesting and approachable. People are left being intimidated by science. But dinosaurs are special—people of all ages love them. So dinosaurs foster a great opportunity to teach science to everyone by tapping into something everyone is interested in.

That's why YANG Yang and ZHAO Chuang are so important. Both are extraordinarily talented, very smart, but neither are scientists. Instead they use art and words as a medium to introduce dinosaur science to everyone from small children to grandparents—and even to scientists working in other fields!

ZHAO Chuang's paintings, sculptures, drawings, and films are state-of-the-art representations of how these fantastic animals looked and behaved. They are drawn from the latest discoveries and his close collaboration with leading paleontologists. YANG Yang's writing is more than mere description. Instead she weaves stories through the narrative or makes the descriptions engaging and humorous. The subjects are so approachable that her stories can be read to small children, and young readers can discover these animals and explore science on their own. Through our fascination with dinosaurs, important concepts of geology, biology, and evolution are learned in a fun way. ZHAO Chuang and YANG Yang are the world's best, and it is an honor to work with them.

Return to the Distant Age of Dinosaurs

In recounting Earth's evolutionary history of life, dinosaurs should never be ignored. They ruled Earth for 165 million years, having multiplied in large numbers, developed into numerous species, and spread to all corners of the world. Their unworldly looks and numerous species continue to fascinate us. Sixty-six million years after most dinosaurs left the world, their cold fossils charm us and remind us of their glorious times and tragic ending.

Now, more and more people crave more knowledge about them. We are curious about a group of animals that came from a distant time, and learning more about dinosaurs is a wonderful way to understand about our planet Earth. One fact may surprise you: not all of the amazing dinosaurs have disappeared. Some of them had evolved into birds and are now flying around us. Yes, we have been living with dinosaurs since ancient times. They built a bridge that connects us to the Earth's distant past.

This book, *Age of Dinosaurs*, gives a snapshot of dinosaurs' history. It starts from the birth of dinosaurs and ends with the extinction of all non-avian dinosaurs. We use a 2D-film approach to show readers this age's most important species, as well as the rise and fall of different dinosaur groups, to sketch how dinosaurs evolved. By doing so, we hope to cover all aspects of this age.

It's a book that children can read on their own, and it's also great for parents to read to their children. In writing about the happy, sad, and cruel moments, the authors reached beyond the charms of dinosaurs; we learned about the fickleness of life. To understand Earth again, we have to ditch our human arrogance.

This, we think, is the ultimate reason that we revisit the Age of Dinosaurs.

YANG Yang
Beijing, July 2019

It was a dawn of a blue planet 234 million years ago.

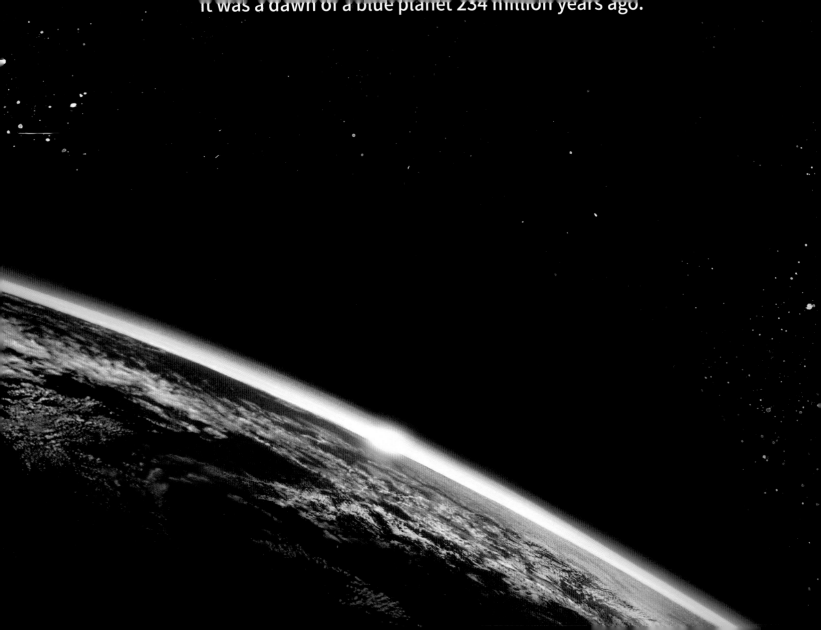

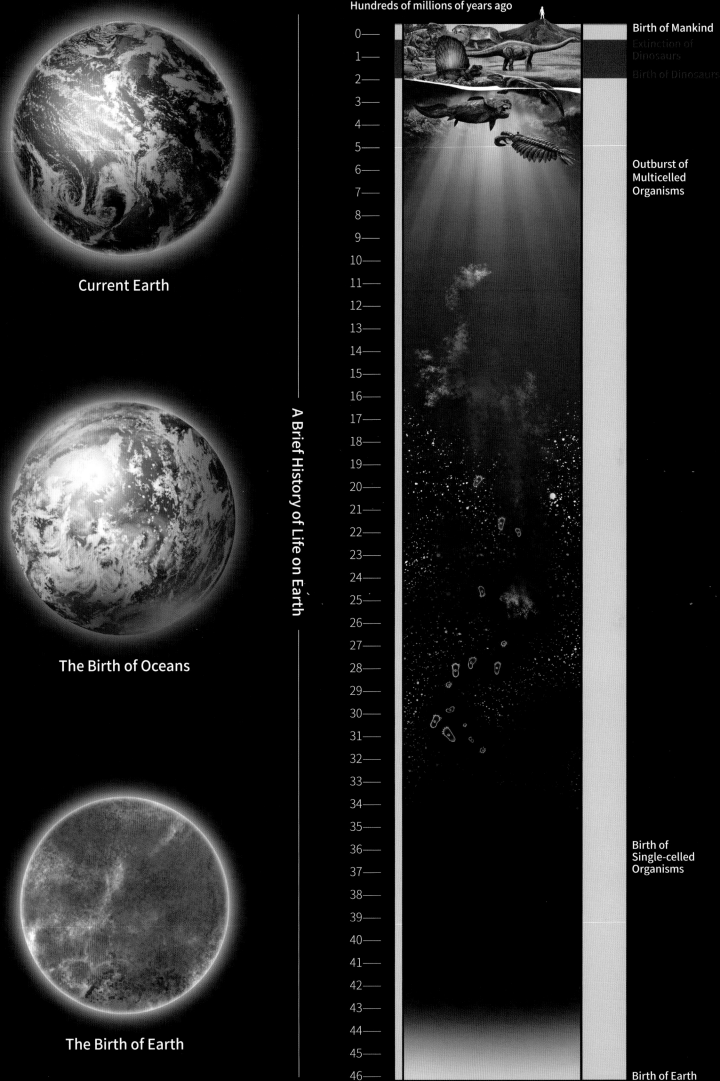

Current Earth

The Birth of Oceans

The Birth of Earth

A Brief History of Life on Earth

Hundreds of millions of years ago

0
1
2
3
4
5
6
7
8
9
10
11
12
13
14
15
16
17
18
19
20
21
22
23
24
25
26
27
28
29
30
31
32
33
34
35
36
37
38
39
40
41
42
43
44
45
46

Birth of Mankind
Extinction of Dinosaurs
Birth of Dinosaurs

Outburst of
Multicelled
Organisms

Birth of
Single-celled
Organisms

Birth of Earth

Geological Timetable of the Dinosaurs Depicted in This Book

Reference: International Chronostratigraphy Chart 2014
Source: International Union of Geological Sciences (IUGS)
Illustrated by: PNSO

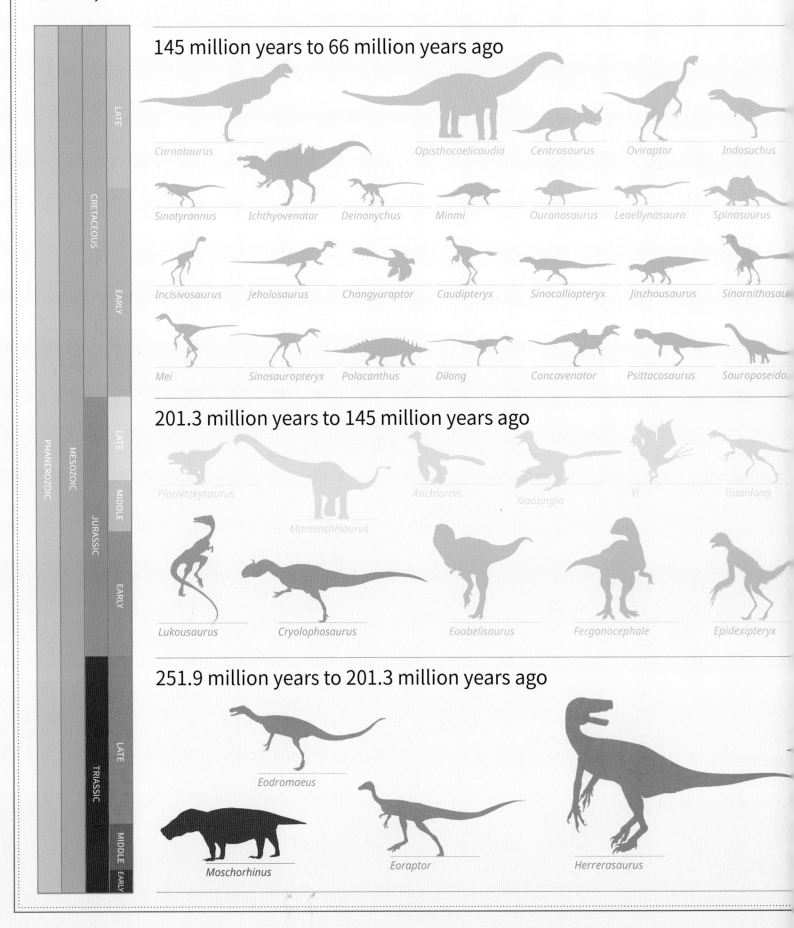

145 million years to 66 million years ago

Carnotaurus · Opisthocoelicaudia · Centrosaurus · Oviraptor · Indosuchus

Sinotyrannus · Ichthyovenator · Deinonychus · Minmi · Ouranosaurus · Leaellynasaura · Spinosaurus

Incisivosaurus · Jeholosaurus · Changyuraptor · Caudipteryx · Sinocalliopteryx · Jinzhousaurus · Sinornithosaurus

Mei · Sinosauropteryx · Polacanthus · Dilong · Concavenator · Psittacosaurus · Sauroposeidon

201.3 million years to 145 million years ago

Piatnitzkysaurus · Mamenchisaurus · Anchiornis · Xiaotingia · Yi · Guanlong

Lukousaurus · Cryolophosaurus · Eoabelisaurus · Ferganocephale · Epidexipteryx

251.9 million years to 201.3 million years ago

Eodromaeus

Moschorhinus · Eoraptor · Herrerasaurus

Nankangia
Euoplocephalus
Triceratops
Tyrannosaurus
Ankylosaurus

Deinocheirus
inoceratops
Pachycephalosaurus
Majungasaurus
Yulong
Qiupalong
Xixiasaurus
Qianzhousaurus

Shuangmiaosaurus
Giganotosaurus
Unenlagia
Velociraptor
Protoceratops
Diabloceratops
Tsintaosaurus

Huaxiagnathus
Yixianosaurus
Sinovenator
Luoyanggia
Siamotyrannus
Yunmenglong
Liaoningosaurus
Ruyangosaurus

Liaoceratops
Amargasaurus
Dongbeititan
Anoplosaurus
Iguanodon
Yutyrannus
Microraptor
Beipiaosaurus

Yinlong
Kentrosaurus
Ceratosaurus
Bothriospondylus
Allosaurus
Europasaurus
Chaoyangsaurus

Tianyulong
Epidendrosaurus
Megalosaurus
Huayangosaurus
Pedopenna
Dryosaurus
Compsognathus

Saturnalia
Plateosaurus
Pterospondylus

Origins of the Dinosaur Fossils Referred To in This Book

Illustrated by: PNSO

ASIA	CHINA	**Present-Day China**
	MONGOLIA	Yi, Liaoceratops, Sinoceratops, Dongbeitian, Microraptor, Pedopenna, Mei
	LAOS	Psittacosaurus, Guanlong, Epidexipteryx, Sinocalliopteryx, Qiupalong, Sinornithosaurus, Anchiornis
	INDIA	Changyuraptor, Yunmenglong, Tianyulong, Liaoningosaurus, Yinlong, Sinotyrannus, Beipiaosaurus
	KYRGYZSTAN	Tsintaosaurus, Ruyangosaurus, Epidendrosaurus, Yulong, Jinzhousaurus, Xixiasaurus, Sinovenator
	THAILAND	Caudipteryx, Lukousaurus, Luoyanggia, Yutyrannus, Huayangosaurus, Sinosauropteryx, Xiaotingia
NORTH AMERICA	AMERICA	**Present-Day America**
	CANADA	Sauroposeidon, Deinonychus, Tyrannosaurus, Allosaurus, Pachycephalosaurus, Dryosaurus
SOUTH AMERICA	ARGENTINA	**Present-Day Argentina**
	BRAZIL	Eodromaeus, Herrerasaurus, Carnotaurus, Eoabelisaurus, Piatnitzkysaurus
EUROPE	ENGLAND	**Present-Day England**
	GERMANY	Megalosaurus, Iguanodon, Polacanthus, Anoplosaurus, Bothriospondylus
	SPAIN	
AUSTRALIA	AUSTRALIA	**Present-Day Australia**
		Minmi, Leaellynasaura
AFRICA	EGYPT	**Present-Day Egypt** — Spinosaurus
	TANZANIA	**Present-Day Tanzania** — Kentrosaurus
	SOUTH AFRICA	
	NIGER	
	MADAGASCAR	
ANTARCTICA	ANTARCTICA	**Present-Day Madagascar** — Cryolophosaurus

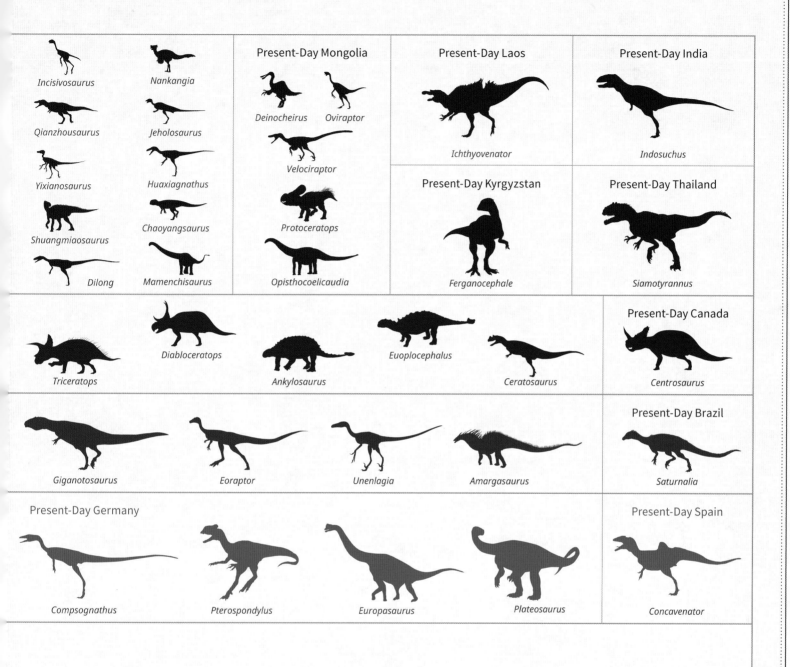

Present-Day Mongolia

Present-Day Laos

Present-Day India

Incisivosaurus

Nankangia

Qianzhousaurus

Jeholosaurus

Yixianosaurus

Huaxiagnathus

Shuangmiaosaurus

Chaoyangsaurus

Dilong

Mamenchisaurus

Deinocheirus

Oviraptor

Velociraptor

Protoceratops

Opisthocoelicaudia

Ichthyovenator

Indosuchus

Present-Day Kyrgyzstan

Present-Day Thailand

Ferganocephale

Siamotyrannus

Triceratops

Diabloceratops

Ankylosaurus

Euoplocephalus

Ceratosaurus

Present-Day Canada

Centrosaurus

Giganotosaurus

Eoraptor

Unenlagia

Amargasaurus

Present-Day Brazil

Saturnalia

Present-Day Germany

Compsognathus

Pterospondylus

Europasaurus

Plateosaurus

Present-Day Spain

Concavenator

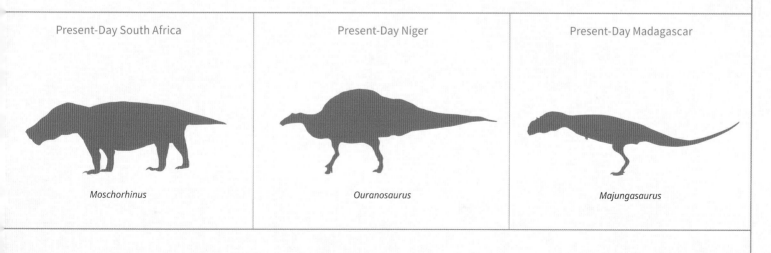

Present-Day South Africa

Present-Day Niger

Present-Day Madagascar

Moschorhinus

Ouranosaurus

Majungasaurus

The Lonesome Triassic Period

From roughly 250 to 200 million years ago, Earth went through the Triassic period.

There was only a single supercontinent, at that time, called Pangaea. A superocean surrounded it.

Moisture from the ocean could not reach far into the Pangaean interior, causing huge, dry, and scorching hot deserts to form at the center of the continent. However other parts of the continent were still mildly warm and humid. These warm and mild temperatures spanned as far as the north and south poles. There were no glaciers at that time.

These climatic conditions favored ferns and conifers. Thus, these plants evolved and flourished.

Thanks to the more accommodating climate, all kinds of life forms started to recover from their past demise in the Permo-Triassic extinction disaster. Dinosaurs began to appear, poising to dominate Earth soon.

However, the dinosaurs were still weak, and dominating the Earth would take time. As the dinosaurs evolved, slowly increasing in numbers and variety, in any part of the world there were only a few of them. Thus, the dinosaurs often wandered alone.

250 Million Years Ago
Present-Day South Africa

The Permo-Triassic extinction destroyed nearly all life on Earth, creating an eerie silence.

Two hundred fifty million years ago, a lucky *Moschorhinus* survivor took a fleeting look at the desolate world under the dim morning light. There were practically no signs of life, but time went on. The Paleozoic Era passed, ushering in the Mesozoic. The *Moschorhinus* stood alone on the land, lonely and fearful.

Luck allowed this group of Therapsids to survive the disaster, and the *Moschorhinus* hoped that luck would favor them enough to allow them to dominate the land. Alas, it did not know that as the recent disaster destroyed the old life forms, it also helped new ones to breed. It was being transformed into something new as the threats of death and extinction continued to whistle and batter against its ears.

234 Million Years Ago
Present-Day Argentina

Finally, 234 million years ago, in present-day Argentina, South America, a reptile stood up on its hind to the surprise of all its friends. For the first time, its large eyes were able to gaze out on the horizon from such an elevated height. Its nasal passages no longer had to be filled with dusty and dirty smells, as it could now enjoy the fresh air found far above the ground. It blinked its eyes closed, reveling in the comfort.

When the *Eodromaeus* began to stand up, it marked the formation of a new species, namely dinosaurs.

Many great things start from humble beginnings. The Therapsids at the time were hoping that luck would shine upon them. However, they failed to foresee that dinosaurs, who were weak and unremarkable at the time, would become kings of the land, eventually ruling Earth for almost 170 million years.

231 Million Years Ago Present-Day Argentina

A *Herrerasaurus*, appearing with its blue stripes, was looking back at a gorgeously colorful sunset. Its charming smile was directed towards its lover who was following close behind.

Two hundred thirty-one million years ago, the *Herrerasaurus* appeared as a new member of the dinosaur family.

Though it was still confined to South America, the birthplace of the dinosaur, it was no longer fragile. The *Herrerasaurus* was big and strong, with a size six times as large as the *Eodromaeus*. Its hind limbs were agile and hardened from the battles it had fought. The *Eodromaeus* had sharp carnivorous claws and teeth. They were building themselves a kingdom of unprecedented scale.

225 Million Years Ago
Present-Day Brazil

Two hundred twenty-five million years ago, a species called *Saturnalia* appeared in present-day Brazil. It was refreshingly new in form, and it was a sign of just how quickly dinosaurs were evolving at the end of the Triassic period.

Distinct from the early meat-loving dinosaurs, *Saturnalia* fed exclusively on plants, making full use of its long neck and tail. But like most herbivorous species, the *Saturnalia* tempted the appetite of carnivores like *Herrerasaurus*. Starting around this time, the dinosaur family was divided into two camps, as the hunters and the hunted fought a protracted battle for survival until their eventual extinction.

214 Million Years Ago
Present-Day Germany

The rule for survival in nature was cruel and paid no consideration to sympathy or etiquette. If one wavered in the face of danger, it meant the loss of power, status, and even one's life. Fortunately, dinosaurs, who had just come into the world, understood this and used this rule to their advantage.

It was 214 million years ago in present-day Germany when a *Plateosaurus* looked back warily. It was one of the earliest plant-eating dinosaurs, and its means of fending off attackers was to grow a bigger body. However, this solution was not as effective as it had hoped as the attackers' sharp claws still found ways to mercilessly puncture through the *Plateosaurus*'s skin.

"Hey! I'm not afraid of the darkness before dawn!" The *Plateosaurus* said to itself encouragingly. To survive, it had to become a little bigger, like a small hill. Until then, it could not let down its guard nor give up hope.

The Bustling Jurassic Period

From about 201 to 145 million years ago, Earth entered the Jurassic period. At the start of the Middle Jurassic, Pangaea began to split.

During this period, the climate became much more humid. Plants evolved and adapted to this change quite rapidly. Gymnosperms, which are plants that produce unenclosed seeds, were at their peak. Ferns, conifers, and cycads covered our planet. There were lush forests everywhere, and most of them featured sequoia, ginkgo, and *Podocarpus* trees.

As flora developed and multiplied, fauna also flourished. Dinosaurs started to live in all parts of the world, growing larger in size and in numbers. They could be found in wet swamps, dense forests, or empty highlands. They gradually began dominating various ecological niches.

190 Million Years Ago
Present-Day China

As the Triassic waned away and the Jurassic arrived, the heyday of the dinosaurs began.

The dinosaur family had grown in diversity and in numbers. The dinosaurs now walked around Earth at their leisurely pace, and love was a thing they could think about in their free time.

It was 190 million years ago, in today's Yunnan Province, China, when the mating season made every dinosaur look attractive.

A beautiful, green-striped female *Lukousaurus* smiled sweetly as it turned around and winked at three male *Lukousaurus* eagerly flaunting their bodies.

The female *Lukousaurus* enjoyed this moment while feeling a little nervous about which one to pick. Only a suitable partner could reproduce successfully. Indeed, finding a mate was not just about following one's heart, as it could decide the fate of the tribe! Contemplating this, the female *Lukousaurus* felt more nervous and uncertain.

The *Lukousaurus* might have belonged to the Ceratosaurs, a branch of the Theropod family. They were the main predators in the early days of dinosaurs. They were on top of the food chain, and that probably explained why they could leisurely pick their mates!

188 Million Years Ago
Present-Day Antarctica

An increasing number of dinosaurs were enjoying life in different parts of the world, as emigrating from South America did not take very long.

It was 188 million years ago in present-day Antarctica, where a *Cryolophosaurus* was about to lay hands on its prey. It failed to notice, however, that it was already in foreign territory. Danger loomed.

Differing from the *Lukousaurus*, the *Cryolophosaurus* belonged to another branch of the Theropod family, the Tetanurae. As this group emerged, tension increased between them and the Ceratosaurs, and thus the battle between the carnivorous dinosaurs began.

168 Million Years Ago
Present-Day China

As new varieties of dinosaurs appeared, the competition between the dinosaurs became ferocious. The dinosaurs that wanted to survive understood the need to innovate.

Night had fallen, but this poor little creature was still hungry. It thought about putting its slender fingers into a tree hollow to catch some tasty worms. However, before trying this out, it had to scout around to make sure no one was trying to eat it!

This little creature, the *Epidendrosaurus*, lived on Earth 168 million years ago in present-day northeast China. It belonged to the Scansoriopterygidae family of the Coelurosauria clade. Although it was far from being able to fly, it could climb trees and hang on them, making use of its long fingers and bent claws.

166 Million Years Ago
Present-Day England

Life produces unexpected events, but they are by no means always pleasant. This *Megalosaurus* had an understanding of that.

One hundred sixty-six million years ago, in present-day England, a *Megalosaurus* was planning for a rear attack against a *Dacentrurus*. The *Megalosaurus* thought its plan was perfect. It had stealthily snuck up behind the *Dacentrurus* without being noticed. It opened its mouth in anticipation as saliva dripped down onto the back of the *Dacentrurus*, and it bit down with the intention of taking a bit of the *Dacentrurus*'s tail. Unexpectedly, however, the *Dacentrurus* swiftly raised up its spiky tail without making a sound and mercilessly smacked the *Megalosaurus* in the face.

"Don't drool over me!" the *Dacentrurus* shouted at the *Megalosaurus* with fury.

The *Megalosaurus* scrambled and ran away, its face still hurting and its mind unable to comprehend what just happened. Of course, it did not know that this seemingly unremarkable herbivore was different from ordinary prey. This herbivore represented a group who could actively fight predators. This was the rise of the *Stegosaurus*, a dinosaur that was brave and could bring hope to the history of the herbivorous dinosaurs!

The *Megalosaurus* would later learn that to be a good hunter, it had to abandon its arrogance and act more cautiously.

165 Million Years Ago
Present-Day China

On a quiet afternoon in Sichuan, China, the sky was as blue and clear as a piece of silk. It was a perfect day to lie in the shade of a tree, thinking about nothing, while keeping one's eyes closed and enjoying the fragrant scent of the air.

It may surprise you to know that the lives of dinosaurs were not always full of fighting. Most of their days were peaceful. A *Huayangosaurus* appeared in this picturesque landscape, happily enjoying the gifts of nature. It was not worried about possible danger because it had powerful weapons!

This *Huayangosaurus* already had fully-developed weapons, typical of stegosaurs. It had towering bone plates on its back and sharp spikes on its shoulders and at the end of its tail, making it a competent warrior in its era.

164 Million Years Ago
Present-Day China

One hundred sixty-four million years ago, the world entered the Middle Jurassic period. At this time, dinosaurs had evolved considerably and were beginning to dominate the Earth.

In present-day Sichuan, China, all forms of life were thriving and vibrant. Herbivorous Sauropods continued to grow bigger in size. Standing in the water, the *Omeisaurus* had a body length of seventeen meters. The body of the *Huayangosaurus* (of the Stegosaur family) was covered with spikes. The *Agilisaurus* was quick and nimble, able to outrun ferocious enemies. Carnivorous dinosaurs had become much stronger and deadlier, and their vast size alone could intimidate prey.

Dinosaurs had firmly secured their dominance over land.

164 Million Years Ago
Present-Day China

A *Pedopenna* with gorgeous blue feathers stood on a tree branch, practicing its flying in Liaoning Province, China.

This wasn't its first time practicing. The other fellows living in same the jungle could often hear the *Pedopenna* flapping its wings. Unfortunately, it had not succeeded once in taking flight.

"One, two, three . . ." Once again the *Pedopenna* mustered up its courage to attempt flight. Suddenly, it heard the loud laughter of another *Pedopenna*. "Ha, you simply will never be able to fly! Why are you wasting your time?"

The first *Pedopenna* did not get angry. It fanned its blue feathers, turned around and replied: "If I did not pursue my dream on a day with such sunny and beautiful weather, I would be wasting my time."

The *Pedopenna* was a member of the Coelurosauria clade. It was very much like a modern bird. Although it had feathers, it never did learn how to fly. Still, why should anyone laugh at a dreamer?

163 Million Years Ago
Present-Day Argentina

On an early morning in Argentina, there was a downpour in the forest, and everything was washed clean.

Shining through the quiet forest, the sun brought a little warmth to the flora and fauna there. Most of the animals were still in a sweet slumber, but the fierce *Piatnitzkysaurus* of the Tetanurae family had already awoken. It ambled through the woods, enjoying the fresh air.

The *Piatnitzkysaurus* soon met a *Herbstosaurus*, who also regularly woke early and was flying around. The *Piatnitzkysaurus* raised its head and made a friendly gesture.

They both resided in the jungle. Though each was powerful and strong, they were eventually able to live in peace.

They were strongmen in different worlds, and their paths seldom intersected. However, an occasional meeting like this was good, as it allowed them to share a little something so their otherwise solitary lives would be less lonely.

160 Million Years Ago
Present-Day China

The seemingly unrelenting sun eventually backed off, and a scarlet orb sank on the horizon. Next, there were dense shades of deep purple and pastel blue in the sky.

Everything appeared to be soft but sturdy, even the large cumulonimbus clouds, which were hanging a dozen kilometers above the ground and looked like giant pieces of Jell-O suspended in the sky.

A rainy evening was approaching. The noisy jungle would soon quiet down.

This was the happiest moment for the animals. Their steps became light and easy. They carried the food that they had gathered over the course of the day, and they prepared to spend a lovely evening with their families.

This *Mamenchisaurus*, walking in the sunset, lived 160 million years ago in the present-day Chinese province of Sichuan. Although it was a herbivorous dinosaur, no one dared to come close to it because of its huge body.

160 Million Years Ago
Present-Day China

The darkness began to fade, and the greyish sky was getting brighter. Over the distant horizon, colors could be seen in the sky, slowly moving towards the forest. A male *Anchiornis* held its breath, opening its eyes wide and waiting for the first ray of sunshine to shoot through the foliage.

Soon, the sun came out. The *Anchiornis* excitedly woke up its lover, and both headed toward the forest, which by now was covered in sunshine.

"Slow down! I can't catch up with you!" its lover called gently. It turned around and smiled softly, "Take your time, I am running ahead of you to make sure that the path is safe."

It was 160 million years ago, in present-day Liaoning Province, China, where a deep love formed between the *Anchiornis* pair as the duo smiled under the morning sunshine.

Anchiornis belonged to the *Troodontidae* family of the *Coeluridae* group. Its body had feathers, but it probably could not fly.

160 Million Years Ago
Present-Day China

In Liaoning Province, China, a pair of *Xiaotingia* stood on a branch, looking into the distance. It was their child's first day to hunt alone, and their hearts remained unsettled. They were afraid that their child might encounter danger or might not get enough food.

The *Xiaotingia*'s child should have been hunting on its own long before this day. However, the *Xiaotingia*'s parents always felt that the child was reluctant to leave. It was on the day of a rainstorm when the father *Xiaotingia* was rescued by its child that the parents finally understood that the child was ready to leave.

In the end, children must leave to face their own challenges. The mother *Xiaotingia* knew that the time was right. She knew that this was the best way to support their child.

The *Xiaotingia* and *Anchiornis* both belonged to the *Troodontidae* family. Its hind limbs had similar scythe-like sharp claws, which were enough to make their prey tremble in fear. Although they weren't feeble creatures, they still had to be cautious as their environment had become very dangerous.

160 Million Years Ago
Present-Day China

While the feathered *Anchiornis* had not realized its dream of flight, the *Yi* with a wing membrane structure was already gliding freely through the forest. The *Yi* was a dinosaur which lived 160 million years ago. Its wing membrane made it unique among all the other dinosaurs that wished to take flight. The wing membrane was similar to that of a present-day bat or flying squirrel. The dinosaurs, imagination for flight far surpassed that of humans. Eventually, the bravest and the most innovative would succeed!

For now, let us look at how this delightful *Yi* couple glided together through the forest. Looking at how they pursued their dreams should encourage us to pursue ours!

160 Million Years Ago
Present-Day China

In Xinjiang, China, a *Guanlong* was sprinting through the forest, chasing after a *Limusaurus*. They had already passed through an open clearing when both dinosaurs starting panting from exhaustion. Despite feeling tired, the *Guanlong* kept up in hot pursuit.

These predators enjoyed showing off their luxuriant fur and bright red crest in front of their lovers, but they were not at all gentle with their prey.

This was not at all surprising. This member of Coelurosauria would have descendants that would become the ultimate over-lords of the Mesozoic Era, namely the *Tyrannosaurus rex*. It was only natural that the ancestors of the *T. rex* would want to show off their muscles.

158 Million Years Ago
Present-Day China

One hundred fifty-eight million years ago, in present-day Liaoning Province, China, a group of small dinosaurs appeared. These cute creatures were only 0.7 meters long with bodies covered with beautiful hair-like feathers. They were not members of Theropoda, which also had feathers, but a separate herbivorous group known as the *Ornithischia* family.

These small creatures with feathers were known as the *Tianyulong*.

Night had fallen, and the moon hung high in the sky, only shining a little light on a few spots on the ground.

The weather was excellent on this evening for the *Tianyulong*. The little creature quietly moved forward in the darkness, dragging its tail, which was far longer in length than its body.

It had to find some food, as its empty stomach had been roaring angrily all day. In those days, it usually went out in the evenings because the sun would shine on its long filamentous feathers in the day, making it dangerously exposed to carnivorous dinosaurs. Because of that, it preferred to be out in the colder but safer moonlight.

155 Million Years Ago
Present-Day England

The day was gloomy, and it was about to rain soon. A gigantic *Bothriospondylus* had not yet satisfied itself with food, but it had to leave now to find shelter. As it moved, it heard something above its head. Turning around, it saw a *Gnathosaurus* passing by.

As a Sauropod, the *Bothriospondylus* had a huge body, a long neck, and a long, thick tail. Most Sauropods could not lift their necks, but the *Bothriospondylus* had longer forelimbs than hind limbs, so it could lift its. That explains how it saw the fast-flying *Gnathosaurus*.

155 Million Years Ago
Present-Day America

Two *Allosaurus*, each about nine meters long, went hunting together.

This was a rare sight. The *Allosaurus* was practically the mightiest dinosaur of the Jurassic. This large carnivorous dinosaur was at the top of the food chain, and it preferred to be alone in hunting, sleeping, walking, or even playing. These creatures were so strong that it seemed that they needed no friends.

However, these were exceptional times. Hunting giant prey required teamwork, and the two of them were doing just that. They were tracing a group of migrating Sauropods, preparing to ambush an old one that had trouble walking.

Soon, the target had completely lost contact with its herd. This was the opening they were waiting for. With their sharp teeth and claws ready, they would rush in and bite its neck and kill it in seconds. The catch could feed them for a week.

155 Million Years Ago
Present-Day America

A *Mesadactylus* and a *Comodactylus* swooped down from a high altitude toward a school of fish in the pond, with their large wings flapping furiously. A *Ceratosaurus* was leisurely strolling on the land, appreciating the blue sky and warmth. The flapping sound scared the *Ceratosaurus*, and it quickly moved into a defensive position. Luckily, the two pterosaurs were attacking another creature, and the *Ceratosaurus* was not in danger.

Such scenes were not common for the *Ceratosaurus* because they had been great fighters and rarely got to watch others fight. A *Ceratosaurus* had a relatively large body, quick movement, sharp claws, serrated teeth, and a terrifying horn on its nose. They flourished because of their great fighting spirit, which spread to other creatures all over Earth.

154 Million Years Ago
Present-Day Germany

A thick tree trunk fell after being soaked in the rain. A *Europasaurus* nearby was awoken and shocked to find that the trunk had fallen so nearby. The *Europasaurus* had almost been killed it in its dream.

The *Europasaurus* calmed down and wanted to see what happened. It took some courage to place its claws on the trunk. With a little push a miracle happened—it stood up!

The *Europasaurus* was very excited, so much so that it stretched its neck almost straight. It had never done this before.

This was the first time that it could see so far into the distance. The view was much better than one blocked by the countless roots lining the shadowy jungle. From the top of the trunk, the *Europasaurus* could see an entirely new world. It took a deep breath. Even the air smelled fresher.

It sat on the tree trunk, looking excitedly into the distance. It saw greener foliage and a curious new world ahead.

The *Europasaurus* decided to go and explore this faraway place, not only for the food it saw there but also to satisfy its curiosity.

73

150 Million Years Ago
Present-Day China

In present-day Liaoning, China, a *Chaoyangsaurus* was chased by a carnivorous dinosaur into the forest. The *Chaoyangsaurus* stopped behind a sturdy tree trunk. It looked around, trying to find an ideal hiding place.

The *Chaoyangsaurus* was composed. It was a forerunner of the *Ceratopsidae* family, the family would eventually evolve to be better defended against predators. It had no sharp horns yet, but it already knew when to hide and when to fight.

150 Million Years Ago
Present-Day America

As night fell, the bustling forest became quiet. Many animals already did their day's work of finding food, but the industrious *Dryosaurus* had not finished yet. Its eyesight was excellent even in dim light. It could find the most preferred leaves in the moonlight. It supported its body with one hind leg, moving its mouth closer and closer to the leaves. Yet it was not completely obsessed with finding food. It kept its muscles taught, ready to react if an enemy showed up.

The *Dryosaurus* was without the armor of the *Stegosaurus* or the gigantic body of the Sauropods, but it could run at up to forty kilometers per hour! As soon as it perceived a threat, it would accelerate immediately, outrunning most of its enemies.

150 Million Years Ago
Present-Day Germany

Rather than expending huge effort trying to become the same as others, it is better to accept your own identity because that is who you really are. For example, the *Compsognathus* of the Coelurosauria clade never expected to have the body of an *Allosaurus*. It would have preferred to be elegant and small.

The morning mist had yet to disperse as the tree trunks were still covered with dew. It was about 150 million years ago, in present-day Germany.

An industrious dragonfly flew over a *Compsognathus* that was in deep sleep. The thin wings accidentally rubbed against the *Compsognathus*'s nose. The small dragonfly was terrified and wanted to flee. The body length of the *Compsognathus* was only about one meter, a length that was small for a dinosaur but huge to a dragonfly. It was too late—the *Compsognathus* turned, stood on the tree trunk, raised its beautiful lower jaw, and in one swift motion caught the dragonfly in its mouth. It would swallow the delicious dragonfly before continuing its dreams.

The Desolate Cretaceous Period

From 145 to 66 million years ago, Earth was in the Cretaceous period, which was the dinosaurs' final era.

Pangaea continued to divide, and Earth looked more like what it is today. However, the continents were not in the positions that they are in today.

Since the last stage of the Late Jurassic, Earth had been cooling, and it was getting even colder in the Cretaceous period. Snowfall increased at high latitudes while tropical regions became more humid than the Triassic and Jurassic periods. Even so, perhaps because of large volcanic eruptions, temperatures began to rise from the Late Barremian onwards.

Dinosaurs still dominated the land and expected to continue their glorious times. Unexpectedly, a disaster lasting a few million years wiped out all non-avian dinosaurs.

This calamity ended the nearly 170-million-year-long rule of the land dinosaurs. Fortunately, life on earth continued to develop despite the dinosaurs' disappearance. New rulers, the mammals, were now emerging from the shadows with eagerness.

The old world had passed away, and a new one emerged. In the midst of desolation, hope was rekindled.

140 Million Years Ago
Present-Day China

82

The best way for Theropods to maintain their hegemony was to grow bigger. However, some in the group did not want to dominate others. These ones did not enjoy hunting bigger prey and preferred to eat small lizards and insects. Thus, their body size remained small. In the golden age of dinosaur evolution that was the Cretaceous period, the number of small and fast-running dinosaurs increased.

The *Mei* curled up, buried its head under forelegs, and fell asleep. Its lovely slumber seemed to get the attention of the moon, which enveloped it in an aura. It was a tiny dinosaur, but its golden feathers gleamed in the moonlight, lighting up the dark night!

Mei belongs to the *Troodontidae* family. Closely related to birds, it had bird-like feathers and slept like birds do.

140 Million Years Ago
Present-Day China

The petite *Sinosauropteryx* fought for lovers gently and elegantly. This did not mean that they were afraid of fighting, but they knew that violence damaged both parties. Thus, the *Sinosauropteryx* would rather avoid it. The *Sinosauropteryx* preferred quiet confrontation, in which the ultimate winner would be the one that persevered until the very end.

One hundred forty million years ago, in present-day Liaoning, China, two male *Sinosauropteryx* confronted one another while competing for a mate. It seemed that they had been standing there staring at each other for a long time. Their beautiful feathers were covered with the mist of early morning.

The standoff appeared calm, but it was filled with murderous tension. The two *Sinosauropteryx* were trying to stay calm, feigning an air of nonchalance to fool the enemy. Their facial expressions might look a certain way, but their body language said something else. Their legs were straight and tense, and they had raised their heads high, revealing the blood-filled skin beneath their necks. They held up their tails in the air, showing off the white rings on their plumage. Their feathers stood up, sending a message that they would not back off.

132 Million Years Ago Present-Day England

Throughout evolution, the defense systems of herbivorous dinosaurs underwent a number of significant changes. Members of the Stegosaur family had spines and plates lining both their backs and tails. From 160 to 110 million years ago, the Stegosaurs were able to defend against predators, but they became extinct after that. Fortunately, their close relatives, the ankylosaurs, lived on to the end of the Cretaceous period. This group made full use of the *Stegosaurus*'s weapons.

First appearing 132 million years ago in what is now England, the *Hylaeosaurus* developed a complete set of armor.

A small carnivorous dinosaur stood in front of it. It warily observed its opponent as its eyes betrayed no sign of panic. The carnivorous dinosaur growled impatiently, hoping for a quick kill to fill its empty stomach. The *Hylaeosaurus*'s composure made the predator more angry than hungry.

War often punished those who were either arrogant or impatient. When the attacker opened its enormous mouth to take a bite, its teeth did not catch any soft meat but hit the swinging spikes of the *Hylaeosaurus*.

130 Million Years Ago
Present-Day China

As the dinosaurs entered their golden age evolution, one group emerged particularly fast. They were puny, but they would flourish and become some of the largest predators on land. Of course, I am talking about the well-known family of Tyrannosaurs.

They would become a well-trained special force unit. With their physical strength, almost nothing stood in their way and they became dominant. By looking at *Dilong*, a feathered and an earliest member of the *Tyrannosauridae* family that lived in present-day Liaoning Province, China, 130 million years ago, we got a glimpse of their strength and ambition.

Compared to *Guanlong*, an earlier Tyrannosaur, *Dilong* was more overbearing and aggressive in appearance.

130 Million Years Ago
Present-Day Spain

A beautiful outlook by no means guaranteed survival. Sometimes it could attract unexpected trouble.

One hundred thirty million years ago, in present-day Spain, a *Concavenator* covered in gorgeous feathers stealthily moved forward to its prey. It was a colorful and attractive creature but often found itself in trouble. The gorgeous red patterns on a protrusion along its back made it easy for it to be spotted by its prey. Thus, the prey often ran away. With a body length of approximately six meters, the *Concavenator* was not small for a carnivorous dinosaur, but its stunning beauty actually made hunting more difficult.

130 Million Years Ago
Present-Day China

In evolution, sometimes number means everything. Those species that have the most offspring like the *Psittacosaurus* would often become dominant.

In present-day Liaoning, China, a mother *Psittacosaurus* led the young out of their burrow to enjoy the warm morning sun as usual. The children were very happy and chirped incessantly. They enjoyed walking around in the forest, breathing in the fresh air, and they liked the crisp sound of their feet stepping on the soft leaves!

Suddenly, the mother *Psittacosaurus* realized that she had lost sight of two of her children. She looked around in panic, only to find that the two little ones were playing hide-and-seek behind a stout tree trunk.

The mother *Psittacosaurus* expressed her anger before hugging her two children. Looking after so many children was not easy. *Psittacosaurus* was a primitive ceratopsid known by its parrot-shaped beak. This small dinosaur was about two meters long.

130 Million Years Ago
Present-Day America

It was an unusually dry season, and the plants were not sprouting new buds. Nevertheless, a *Sauroposeidon*, which lived 130 million years ago in present-day America, was in good spirits.

It was a member of the Sauropoda clade. Not only did it have a huge body, but it was also one of the world's longest dinosaurs. Because of its long neck, it could eat the leaves in high places, which sustained them for a long time.

Other herbivorous dinosaurs were less lucky. At its feet were three famished *Camptosaurus*, which were waiting to see if the *Sauroposeidon* would spare them some leftover leaves.

The *Sauroposeidon* could see them, but it had not decided whether or not to share the leaves with its neighbors. A brief moment of feeling someone's envy could be enjoyable.

The survival strategy of the *Sauroposeidon* was to use their large size to repel all threats. However, this also created difficulties. Their huge body made them clumsy, and when natural disasters struck, they were unable to flee quickly. Unfortunately, when things were going smoothly, it was hard to pay attention to the possibility of a precipitous demise.

Such was the fate of the gleeful *Sauroposeidon*. As North America entered the Cretaceous period, the number of Sauropods drastically fell. The cruel conditions spelled the final chapter of the North American dinosaurs. If our *Sauroposeidon* could foresee the future, would it want to share its last waning moments of happiness with the begging *Camptosaurus*?

130 Million Years Ago
Present-Day China

One hundred thirty million years ago, in present-day Liaoning, China, a *Liaoceratops* stealthily crept toward a fresh fern.

Next to the ferns, a carnivorous dinosaur was sleeping. Although *Liaoceratops* was a little nervous, it inched closer towards its favorite food little by little.

There are other similar ferns in this enormous forest; however, the *Liaoceratops* had an appetite for risk , and it wanted to test its own courage.

Although the *Liaoceratops* had a little crest on its head, it would be of little use if carnivores attacked.

130 Million Years Ago
Present-Day Argentina

Two *Amargasaurus* were foraging in an open plain.

Among Sauropods, the ten-meter-long *Amargasaurus* was small, only as long as a *Hadrosaurus*. But it looked quite different among the family, because of its tall spikes on its back. These spikes were tallest on its neck and shortest on its hip. They formed two rows on its neck but merged into one in the middle of the back. The tallest spike could grow to sixty-five centimeters. Other Sauropods, such as *Diplodocus*, had spikes, too, but theirs were not as tall.

The tall spikes made up for *Amargasaurus*'s small size, helping them to deter predators.

128 Million Years Ago Present-Day China

A *Dongbeititan* walked on the parched soil near a volcano, which had become dormant only recently. Hot lava no longer gushed out, but the air still smelt pungent in present-day Liaoning Province, China.

The *Dongbeititan* was a large Sauropodomorph. Most predators that lived around it were smaller, so it seldom worried about safety. It started to worry only recently when volcanoes were becoming active and killing many. Its size would not make it safer in an eruption, which would be a serious challenge to big and small animals alike.

127 Million Years Ago
Present-Day England

The morning weather was always marvelous in present-day England, and the armored *Anoplosaurus* came to the lake early to enjoy some refreshing water.

But the planned pleasant time was disturbed by a flock of *Caulkicephalus*. They set their sights on those fish leaping out of the water, and they brandished their protruding teeth, ready to pounce.

The *Anoplosaurus* was somewhat disappointed and got ready to leave. It could do nothing to save those poor fish, and the sight of slaughter would make the lake no longer a pleasant place.

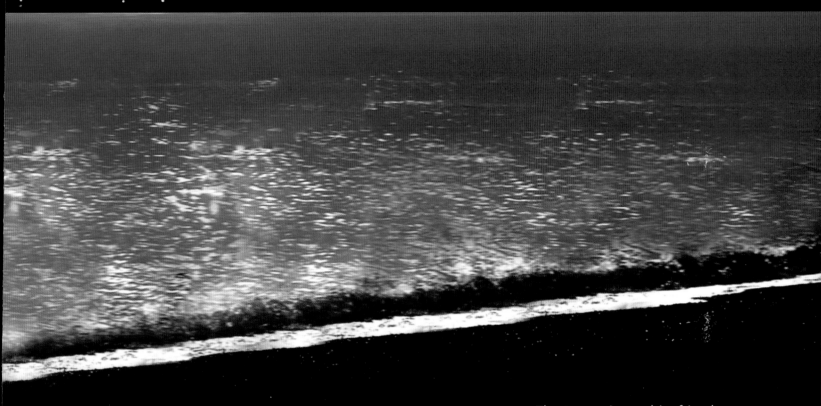

It was a very beautiful sight: the *Iguanodon* reflected as it ambled along. The bright moonlight blended the endless sea and the sky into a single blur, and the soft sand slipping under its feet still kept the remnants of the day's warmth.

Although night had come, many things appeared brighter than during the day. This was the *Iguanodon'* s favorite moment.

The *Iguanodon* and its friends were a bunch of lazy fellows, who derived great pleasure from getting suntanned, enjoying the sea breeze, and gazing at the stars. In fact, the only thing they detested was conflict. Although their forelimbs were fitted with very sharp claws, they had no interest in using them in a fight.

They passed this carefree temperament on to their descendants, including the various members of the *Hadrosauridae* family. You will know what I am talking about when you meet that group.

125 Million Years Ago
Present-Day China

The weather was hot. A pair of *Yutyrannus* hastily ran past, looking rather uncomfortable in this weather due to the coats of thick hair which covered their entire bodies. They looked very much like a pair of strange-looking brutes in thick-padded jackets.

Behind them, from time to time the sound of laughter could be heard. They were not looking back, but they knew that the laughter came from unkind creatures ridiculing them.

However, this did not affect the mood of the two *Yutryannus*, nor would it stop them from becoming the most powerful creatures in the region. They were members of the Tyrannosaur family, capable of growing up to nine meters in length. More importantly, the heat was subsiding, and the snow would soon appear to mark the beginning of the first winter. It would become bone-chillingly cold and bleak, and only those with a coat of heavy fur would withstand the icy temperatures. Unfortunately, those who knew how to laugh at someone else did not learn how to notice the small changes in the environment.

125 Million Years Ago
Present-Day China

Mist enveloped the land, but there was a navy blue stream jumping around in the densely green forest. Most of the animals were in a deep slumber, and everything was quiet. A small lizard that had woken early was clinging to a tree. Its large eyes were observing the things around it. Aside from the lush greenery, it found nothing unusual. Suddenly, a navy blue *Microraptor* lept from the canopy and broke the silence. It stretched its four wings and floated down like a leaf towards the lizards. Its feathered hind limbs were stretched with its claws facing forward. Its partner followed just behind it, even though attacking a lizard was not dangerous at all. The two *Microraptor* lovers enjoyed each other's company at all times, and they felt the warmth of companionship at this moment.

The *Microraptor* lived 125 million years ago in present-day Liaoning, China, and it was partially able to fly. Although its flying skills probably never progressed beyond gliding, it could enjoy the sweet, delectable feeling of gliding down from high altitudes.

125 Million Years Ago
Present-Day China

In present-day Liaoning, China, the lush forests were comprised of broad Japanese cedars and towering sabino. One hundred twenty-five million years ago, the first kinds of flowering plants, *Archaefructus liaoningensis* and *Archaefructus sinensis*, bore their flowers in the shallow ponds. In the treetops, the beautiful *Confuciusornis* and *Liaoningornis* were building their nests. Below, in the bushes, a *Sinosauropteryx* was busily hunting a *Zhangheotherium*, which was the size of a mouse. Several *Caudipteryx* were on the open ground near the lake, ruffling their beautiful feathers to attract mates. In the distance, a *Sinornithosaurus* was sprinting along,

waving its forelimbs, desperately trying to leap up high enough to catch a *Confuciusornis*. Schools of *Lycoptera* and *Perpiaosteus* swam in the lake, enjoying their unimpeded lives. An *Aeschnidium* stood quietly on a branch of deadwood, watching a *Manchurochelys* climbing clumsily onto the bank to bask in the warm sun. A *Callobatrachus*, its hind legs tense, was ready to pounce upon a *Protonemestrius* to get its next meal. At the deep end of the forest, a giant *Dongbeititan* wandered home alone through the bustling noisy bushes.

This was the Early Cretaceous period when dinosaurs entered their golden age.

125 Million Years Ago
Present-Day China

Life is fair to everyone. When a door is shut, a window opens. The cautious *Beipiaosaurus* lost out on a lot of fun as it shunned adventure, but it led a stable and safe life.

A *Beipiaosaurus* was galloping through the woods when it heard a rustling sound coming from behind. It stopped, looking back nervously.

The *Beipiaosaurus* was extremely timid, despite being a member of the powerful *Therizinosauridae* family. All Therizinosaurids had large, sickle-like claws, but the *Beipiaosaurus* disliked both hunting and meat, and eventually became herbivorous or omnivorous, using its claws to gather plants. When needed, it could fight predators with its sharp claws. Its defense was usually enough to deter many attackers, but it always acted cautiously, looking around if it heard something.

Beipiaosaurus looked back for a long time and found no danger. It finally felt reassured and continued to walk into the sunlight.

125 Million Years Ago
Present-Day China

An *Incisivosaurus* looked warily around before quickly entering the woods. Its favorite leaves were by a nearby lake. However, in order to get there, it had to go through the perilous woods.

The soil was moist from a morning rain, but the *Incisivosaurus* had no time to indulge in the fresh air. Its only wish was to get through the forest safely.

The lichens at its feet were absorbing the rainwater, and they appeared smooth and translucent. However, the *Incisivosaurus* stepping over them refused to take one look at them as it ran by quickly.

Suddenly, the *Incisivosaurus*'s foot slipped, its body slanted, and it almost fell down. It screamed in fear with its heart pumping wildly.

The *Incisivosaurus* looked around cautiously. Fortunately, no one seemed to be taking notice of it. It breathed a sigh of relief. It arrived at the lake after no more incidents and checked how it looked in the water, panting desperately.

"Well . . ." the *Incisivosaurus* sighed. It was unhappy about its seemingly meaningless life. It was on pins and needles all day, not even daring to linger a bit longer on the road where it had to pass through every day. It was not a large dinosaur, but it was the founding member of the Oviraptorosaurs. It felt that it should not continue to live in fear.

The *Incisivosaurus* finally made a resolution. From tomorrow onwards, it would never go through the woods in fear. It should enjoy the beauty of the forest, and if something dangerous appeared, it would be able to fight back!

124 Million Years Ago
Present-Day China

The world of the dinosaurs was not always filled with fighting. Most of the time they had nothing much to do and certainly did not fight. It could be quite a task to look for something fun to do!

It was 124 million years ago, in present-day Liaoning, China, where a beautiful *Caudipteryx*, roughly the size of a peacock, found a comfortable spot to spend the day. Aged vines hung down from above. The hot sun was blocked by tall trees, and the *Caudipteryx* felt that nowhere else could be better.

It gracefully fanned out its beautiful tail feathers, trying to make a pose while facing left, and then turning around to try another. This would be a good way to kill time. If a female *Caudipteryx* were to pass by and see its graceful posture, that would be even better.

124 Million Years Ago Present-Day China

A complicated environment required accurate judgment. In order to win and survive, one had to distinguish truth from illusion. Of course, this was not easy.

A *Sinocalliopteryx* stood quietly close to a river. It lifted one foot, tentatively extending it into the water. Letting the gentle flow of water stroke its foot was a very pleasant experience.

Other dinosaurs tended to be captivated by the elegance of the *Sinocalliopteryx*. The more restless ones thought that they had encountered a beautiful princess, and could not wait to approach and flirt. This was perfect for the *Sinocalliopteryx*. As the largest member of the *Compsognathidae* family, it was not as good-natured as the smaller fellows, and it would mercilessly prey upon those who presented themselves up as potential meals.

122 Million Years Ago
Present-Day China

The *Jinzhousaurus* were a group of gentle creatures, tired of the ceaseless fighting in the jungle, and determined to live a merry life.

They were Hadrosaurs, and their children and grandchildren would spread throughout the world.

This was one of the first members of their family, the *Jinzhousaurus*, which lived 122 million years ago, in present-day Liaoning, China.

Looking at their optimistic character, harmless mouths, and their innocent expression, it is apparent that they really wanted to live an idyllic life of leisure.

122 Million Years Ago
Present-Day China

Life would sometimes go off track. Even though the *Jinzhousaurus* would love to go into hiding and escape from everything, there would always be something that pulled it back to reality. This time, it was probably facing life's greatest challenge.

The three-meter-long *Jinzhousaurus* had to fight three attacking *Sinornithosaurus*, each one 0.7 meters in length.

Fighting in the open sunshine did not bode well for the *Jinzhousaurus* as it could not run for cover. The *Sinornithosaurus*'s hind legs were equipped with curved, sharp claws that could cut deep into an enemy's skin. Their sharp teeth would soon pierce the *Jinzhousaurus*'s skin. It was a one-sided battle, and the best thing the *Jinzhousaurus* could hope for was to lose the fight and run away with new wounds.

122 Million Years Ago
Present-Day China

Like the *Sinocalliopteryx*, the *Huaxiagnathus* was also a large-sized member of the *Compsognathidae* family, but it did not like to play games with its emotions always written on its face. There was nothing wrong with this, but it could not hope for prey to come looking for it. Instead, it had to make efforts to catch anything it could.

A *Huaxiagnathus* walked along a steep trail up the hill, and the gravel underneath was unkind to its feet. It was not climbing the hill to view the scenery, but to hunt for prey.

How could prey appear on top of such a barren hill?

Well, indeed they would come! Most hunters would think that prey might not climb up this hill, and precisely because of that, this place had become a sanctuary for many herbivorous dinosaurs.

The intelligent *Huaxiagnathus* discovered this secret long ago, so even if it had blistered feet from climbing up the hill, it still made it a habit to visit the hilltop once a week, and it had never been disappointed.

122 Million Years Ago
Present-Day China

All that the *Yixianosaurus* wanted was one tall tree, yet even that was difficult. The *Yixianosaurus* was complaining just yesterday about the towering trees blocking its sight, but the whole world abruptly turned dim overnight.

It lost its dear child in a battle when it was ambushed.

The *Yixianosaurus* had no memory of how it spent its time since last night. It probably had walked a long distance, away from that battleground, shedding tears on the way. At the moment, it was looking for a tree to stand on, just like what it used to do with its child.

It was likely that the *Yixianosaurus* was related to the *Epidendrosaurus*. Like the *Epidendrosaurus*, it too had claws which were well suited to climb trees. Its best memories had to be the ones of time spent on trees.

122 Million Years Ago
Present-Day China

The *Sinovenator* was only about a meter long. Although it had feathers, many other dinosaurs that lived in the Early Cretaceous period in present-day Liaoning, China, were also endowed with beautiful long feathers. At first glance, it did not appear to be special. However, the first impressions were not enough to describe this truly brilliant star.

Observing its incredibly long hind legs and its feathered forelimbs, it is apparent that its mode of locomotion was very different
from that of other dinosaurs. Its forelimbs, much like a bird's, were
able to stretch out to the sides. It was an important evolutionary
landmark between typical dinosaurs and the avian species. And
that was what made this creature special 122 million years ago.

120 Million Years Ago
Present-Day China

A layer of fallen leaves blanketed the ground. Moving gracefully over them was a beautiful bird-like dinosaur, and its body was entirely covered in feathers. It walked gracefully, moving its slender legs. Suddenly, it detected a curious-looking insect on the ground. It stopped, stretching its forelimbs, and tiptoed toward to the insect. However, the mild fluttering of its forelimbs disturbed the leaves on the ground, and the insect fled.

The insect disappeared; however, this elegant dinosaur was not too disappointed. It continued its journey; after all, this little insect was only an amusing distraction along its path.

This dinosaur, which lived 120 million years ago in present-day Henan, China, was the *Luoyanggia*. It was a beauty among the *Oviraptors*.

120 Million Years Ago
Present-Day Thailand

At the apex of the dinosaurs' development, the Tyrannosaur group was the most outstanding of all of the carnivorous dinosaur families. Not only did it constantly increase in physical size, but it also expanded in geographical distribution. With an ever-growing number of Asian members, it was gathering strength and beginning to dream of the day when it would conquer faraway territories.

One hundred twenty million years ago, in present-day Thailand, Asia, a *Siamotyrannus* with its large mouth wide open showed its inviolable authority towards its prey and enemies.

Although the *Siamotyrannus* was more primitive than the *Tyrannosaurus rex*, it was more than seven meters in length and wielded great strength. It was worthy of being considered the king of its region.

120 Million Years Ago
Present-Day China

The weather was exquisite, and the sunshine placed a golden hue on everything, including the clouds, mountains, trees, land, and a *Yunmenglong*.

The *Yunmenglong*'s mood was cheerful and satisfied because it had a full stomach.

For a long time, the lush trees supplied it with abundant food, enough to satisfy its immense appetite. Sometimes it ate so much that it found walking difficult.

We now know that 120 million years ago, today's Henan, China, was a paradise for the *Yunmenglong*, an enormous member of the *Sauropoda* family.

120 Million Years Ago
Present-Day China

The evening sun always evoked a sense of mystery. It could paint the sky with bright colors and sprinkle the lake with a mysterious light. When evening came, the world seemed to be cleansed, as the dark green foliage instantly turned to orange, and the white clouds were dyed in the most brilliant of colors. If we look at the world 120 million years ago, apart from the evening sun, few things were the same. These two *Liaoningosaurus*, which lived in present-day Liaoning, China, were spectacular in appearance. These creatures had scales covering their bellies and were fond of swimming and catching fish, but, surprisingly, they belonged to the *Ankylosauria* family.

The *Ankylosauria* family generally had their entire bodies covered in armor except for the abdomen area. Carnivorous dinosaurs knew that was a soft spot to attack. Because of this, the unique *Liaoningosaurus* evolved to grow scales to cover their soft underbelly!

120 Million Years Ago
Present-Day China

The sun was partially covered by clouds, but it was still baking the land dry in present-day Henan, China. A *Ruyangosaurus* of about thirty-eight meters long was slowly approaching a small lake of clear water. Though its head had to endure the scorching hot sun, its tail was feeling a cool breeze. A bolt of lightning split the dark clouds above its tail, foreshadowing a downpour.

Ruyangosaurus was one of the world's largest dinosaurs. Because of its great size, its head and tail could be in two different weather conditions.

It was the hot and stuffy weather that drove the *Ruyangosaurus* to the lake. However, its appearance stirred panic among the *Oviraptors* nesting at the lake. Although the *Ruyangosaurus* was a gentle dinosaur and its favorite food was succulent leaves, it was enormous, such that one careless step would destroy an entire nest of *Oviraptor* hatchlings. The mother *Oviraptors* had to exercise vigilance, guarding their children, whilst screeching to warn the *Ruyangosaurus* not to enter their territory.

**120 Million Years Ago
Present-Day China**

A *Sinotyrannus* ran past a pool on the outskirts of the forest in Liaoning, China. Compared to the petite early members of the Tyrannosaur family, it was mightier with a body length of approximately ten meters. Its mouth was full of sharp-edged fangs that dripped with saliva, and its vocal chords were capable of producing fearless roars.

In front of its gaze, prey was panic-stricken and fled. The *Sinotyrannus* did not deem it worthwhile to accelerate but maintained a steady pace, knowing for certain that very soon it would be dining on this pitiful prey. This was an unchallenging hunt that did not require the *Sinotyrannus* to sacrifice its dignified composure.

116 Million Years Ago
Present-Day Laos

An *Ichthyovenator* was getting ready to catch fish. Because of its intriguing appearance, it was a very conspicuous fellow.

On its back sat a sail, which for some reason sank in the middle, split into two halves. The feature was different from its cousins, and this got the attention of the opposite sex.

The *Ichthyovenator* belonged to the *Spinosauridae* family, a very special group of large carnivorous dinosaurs. It had a head similar to that of a crocodile, and its mouth was equipped with cone-like teeth, which had either no jagged edges or very small ones. While the other carnivorous dinosaurs devoted all their time to capturing larger prey to feed on for days, the *Ichthyovenator* was content to catch fish. The *Spinosauridae* family appeared during the Late Jurassic period, and by the Early Cretaceous period, they had a flourishing population.

115 Million Years Ago
Present-Day America

"Whoa . . ." a group of *Tenontosaurus* rumbled as they roamed through the plains of present-day Montana, United States, 115 million years ago.

Their noticeable movements attracted the attention of a group of resting *Deinonychus*, which grew excited. The *Tenontosaurus* was one of the *Deinonychus*'s favorite foods, and the many prey running past them were enough food for the entire group.

The *Deinonychus* immediately gathered themselves, moved into attacking positions, and got ready to use their sharp, sickle-shaped claws on their forelimbs, which were characteristic of the *Dromaeosauridae* family. The attack began, with the attackers constantly changing formations at amazing speed, searching for the most effective point of attack. Look at how they did it!

115 Million Years Ago
Present-Day Australia

In present-day Australia, a *Minmi* walked along the plains, searching in all directions for low ferns.

The weather had been splendid with plenty of rainfall, yet those ferns were nowhere to be found. The *Minmi* cursed under its breath as it lowered its head to continue its search.

The *Minmi* might be excused for its rude manners. Few could possibly stay in good spirits whilst trudging along on a long journey with an empty stomach.

One good thing that the *Minmi* could be proud of was that it would not be easily eaten by predators. It had the advantages of a typical Ankylosaur, with its entire body shielded with rows of armor plates and thorny spikes. These protections would deter the carnivorous dinosaurs, so its search for food would not be too risky.

110 Million Years Ago
Present-Day Australia

As Earth entered the Cretaceous period, most places on Earth were still warm year-round, but some places began to see snowy winters.

One hundred ten million years ago, in present-day Australia, snowflakes were swirling in the sky, and soon the land would face six months of polar night. Hence, the enormous *Muttaburrasaurus* was preparing to migrate north.

Yet a group of the smaller *Leaellynasaura* would not have enough energy to carry out such a long journey, so they chose to hibernate. However, before going into a long, deep slumber, they had to collect enough food to survive winter.

The *Leaellynasaura* stood under the full force of the icy wind and snow, looking around for plants. The last thing they expected was a three-meter Tyrannosaur leaping from behind a pile of snow, upsetting their nice plans.

110 Million Years Ago
Present-Day Niger

Hundreds of millions of years ago, the world was more complex than we can imagine. It was very difficult to have a peaceful life, and the *Ouranosaurus* evolved to have all sorts of gear so it could defend itself when necessary.

A seemingly tranquil *Ouranosaurus*, with a thick, heavy "sail," walked under the sun. It was a peaceful member of the Hadrosauriformes clade, which always shied away from conflict. However, conflict would sometimes be imposed on it. The sail on its back looked similar to that on a *Spinosaurus*. However, there was one difference. The *Spinosaurus*'s sail grew thinner from its back to its tail, whereas the *Ouranosaurus*'s became progressively thicker. The *Ouranosaurus* did not have the size and strength of the *Spinosaurus*, so in order to make itself look fierce and mighty, it chose to grow its sail to look like a giant.

100 Million Years Ago
Present-Day Egypt

A *Spinosaurus* waded across the stream. Although its movements were very light, it still caused quite a stir.

It was indeed an unusual animal. On its back stood a tall and colored "sail," which moved like a mountain and cast a large shadow. In addition, its size and look had earned it a terrifying reputation among other inhabitants. Although its favorite food was fish, not somebody else's flesh, no one wanted to come near it in case it was feeling adventurous about its diet.

So, while this massive fellow simply wanted to enjoy the cool, refreshing water flow, other inhabitants that lived nearby scrambled for shelter.

Living one hundred million years ago in present-day Africa, the *Spinosaurus* was the representative member of the *Spinosauridae* family and the world's largest carnivorous dinosaur. It could grow to fifteen meters, and the spectacular sail on its back could reach two meters tall. Its limbs were short and had webbed feet. It was a semi-aquatic dinosaur that spent most of its time in water.

2012.6.27

100 Million Years Ago
Present-Day China

It was one hundred million years ago, in present-day Liaoning, China. A lone *Shuangmiaosaurus* gazed despondently into the dark.

It had just fought a fierce battle, and although it made a great effort, it still could not rescue its children from the clutches of a *Yutyrannus*. It could only watch as its children were taken away, with its heart sunken with grief and sorrow. It felt as if it were buried along with its dead children.

In the silence of the night, all of the other animals were in sweet slumber with their families, but this *Shuangmiaosaurus* was struggling to find a reason to live on.

93 Million Years Ago
Present-Day Argentina

A prey sprinted across the forest, attracting the attention of a *Giganotosaurus*. It opened its large mouth, and made a sound, with saliva spitting in all directions. This was a cunning prey, one that tried to hide its body behind the bushes. Yet for the *Giganotosaurus* such tricks were of no use, as it had keen vision.

The Cretaceous period was the pinnacle of dinosaurs' evolution. During this period both herbivorous and carnivorous dinosaurs raced to develop weaponry that would outmatch their enemies. The *Giganotosaurus* used its mammoth size as a deterrent. It was about twelve meters in length, five meters in height, and weighed between six and eight tons. Few enemies could match its enormous size.

Giganotosaurus belonged to the *Carcharodontosauridae* family, one that comprised many large carnivorous dinosaurs; some were almost the same size as the famous *Tyrannosaurus rex*. During the Early- to Mid-Cretaceous period, they ruled over the southern continent together with the *Spinosauridae* family.

90 Million Years Ago
Present-Day Argentina

The weather was oppressively hot, and the moisture from a lake stuck to an *Unenlagia*, gluing its beautiful feathers to its slender body shimmer.

The *Unenlagia* was bothered less by physical discomfort than by its looking less elegant and beautiful. The *Dromaeosauridae* family was renowned for splendid feathers, and it absolutely could not allow its feathers to look so saggy. It angrily shook its body and groomed its feathers meticulously with its mouth. It desperately wished that a cool breeze would soon arrive and blow away the smothering heat and damp.

85 Million Years Ago
Present-Day China

In present-day Zhejiang Province, China, there was a natural habitat for a wide array of herbivorous dinosaurs. During breeding season, the *Dongyangosaurus*, a five-meter-high, 15.6-meter-long Sauropod, would come to the river, seeking loose soil which was suitable for nesting sites. The five-meter-long *Zhejiangosaurus*, although covered with armor plates, quickly dodged out of the way of the enormous *Dongyangosaurus*.

A couple of Hadrosaurs, which made camps near the riverbank at an earlier time, had to watch vigilantly over their eggs in their nest, guarding against threats. After all, the flourishing of the families depended on their successful breeding.

83 Million Years Ago
Present-Day Mongolia

As nightfall came, the world fell into silence. A *Velociraptor* stood alone on a reef close to the shore, enveloped in a feeling of immense loneliness.

Living eighty-three million years ago in present-day Mongolia, this *Velociraptor* never once laid eyes on its parents. When it hatched and laid eyes on this world, it was alone. It did not understand why its parents abandoned it, but it was not looking for a reason. For many years, it was looking for its parents and longing to be with them.

It told its longings to the moon and the sea, hoping that when the sun rose the next day, it would be greeted by the sight of its mother. It would not give up.

83 Million Years Ago
Present-Day Mongolia

A typical *Velociraptor* was not a sentimental creature. It was the perfect hunter on the Mongolian plateau, with its wisdom, three-dimensional vision, keen hearing, and powerful build. It liked to conquer.

It was a hot day, and everyone was feeling lazy, but one *Protoceratops* was diligently looking for food. Its short limbs were moving slowly, a bit too slow in an environment full of danger. As it looked for fresh plants in the blazing sun, an equally diligent *Velociraptor* followed it. When someone more powerful worked more diligently, there was no question about who would come out on top.

Quickly the *Velociraptor* attacked the *Protoceratops* with its sharp tooth, and the fight was soon over. The dust flying around was the only evidence that proved the *Velociraptor*'s fighting strength.

82 Million Years Ago
Present-Day America

The sharp-edged horn of the *Ceratopsia* family did not only act as a weapon.

Living eighty-two million years ago in North America, the *Diabloceratops* was a fashion star of the time. There were more than twenty differently-sized horns growing on the *Diabloceratops*'s face, four of which were particularly long. In addition, its frill was adorned with beautiful and truly fashionable patterns and colors. These adornments helped them to have better chances of attracting a mate.

80 Million Years Ago
Present-Day China

Laying eggs was an important ceremony for dinosaurs. Even the largest of them had to treat this act cautiously.

Every mating season, the dinosaurs living in the vicinity of present-day Xixia, Henan Province, would flock to the wide riverbank to mate and to lay eggs. This place had remained the favorite spot for generations.

The climate of the area was very similar to a subtropical island and populated by crocodiles, turtles, fish, and primitive birds. This precious nesting ground had plenty of sunshine throughout the day, and the ground was relatively flat.

When the dinosaurs laid their eggs, they would first use their claws to press down to make a small, round depression in the ground and then place their eggs around it in a clockwise or counterclockwise pattern. Once the laying dinosaur made a circle of eggs, it would cover them with a layer of leaves and then make one more outer ring, and it would repeat the process until it finished laying all the eggs. This was done to improve the eggs' chances of hatching.

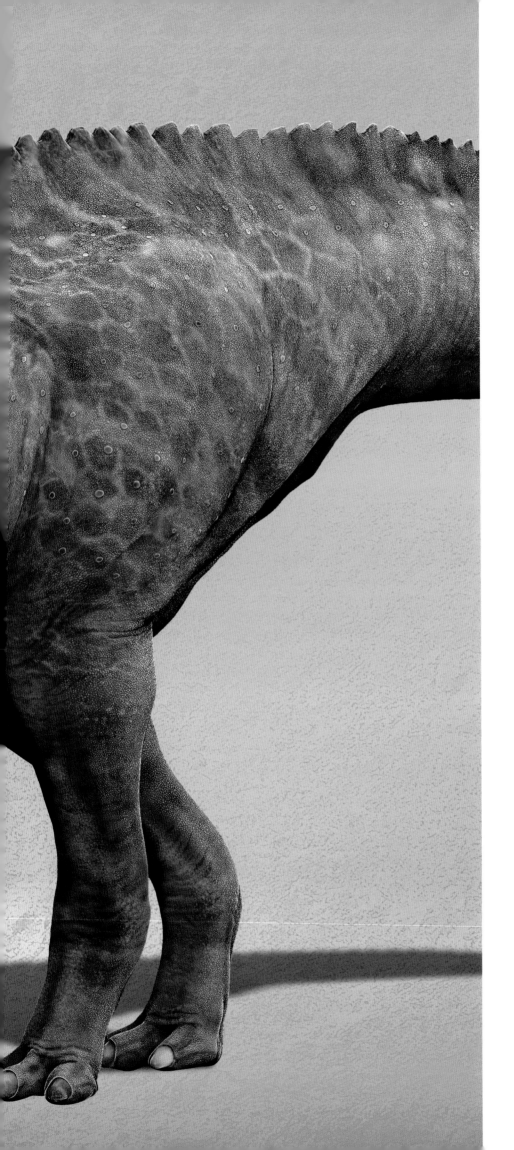

80 Million Years Ago
Present-Day China

In present-day Shandong, China, a *Tsintaosaurus* took a leisurely stroll, its crest swinging in rhythm with the motion of its body. Its mouth had the look of a duck's bill, and it moved with a laid-back demeanor, both suggesting that it belonged to the *Hadrosauridae* family.

The ferns growing on the ground were abundant. The towering ginkgo and pines were marks of the past. The *Tsintaosaurus* took a deep breath to enjoy the flowery scene and moved toward the lake at the edge of the forest.

It was nearly 150 million years after the first dinosaurs appeared. Dinosaurs reached their peak in terms of diversity, and each was thriving in its own respective ecological niche.

80 Million Years Ago
Present-Day China

In present-day Gansu, China, a group of *Hadrosaurus* was enjoying the clear water in the early morning. They lined up with the larger adults encircling the smaller ones to protect them at all times.

This was a common scene for the *Hadrosaurus*. They rarely lived alone. Whether they were gathering food, drinking water, bathing, playing, or even resting, they remained together.

These *Hadrosaurus* had only their huge bodies as a means of defense, so perhaps staying together in a group was the best way to survive. In this way, they grew and bred, spreading across the globe, and becoming one of the most successful dinosaur species of the Mesozoic Era.

75 Million Years Ago
Present-Day Argentina

Soon it would be nightfall in present-day Argentina. Unsuspecting herbivorous dinosaurs were drinking in a river, and this got the attention of a *Carnotaurus*. It was rare for herbivorous dinosaurs to wander alone, and this rare opportunity to strike could not be missed. This *Carnotaurus* had a body nine meters long; its hips reached a height of three meters. It weighed approximately 1.5 tons. At this moment, it tensed its muscles and opened its jaws.

Next to come was a rare scene. The *Carnotaurus* charged towards its prey faster than any other large carnivorous dinosaur.

Carnotaurus was a famous member of the *Abelisauridae* family. The family existed in the Late Cretaceous period, replacing the *Carcharodontosauridae* family as the predators in the southern continents.

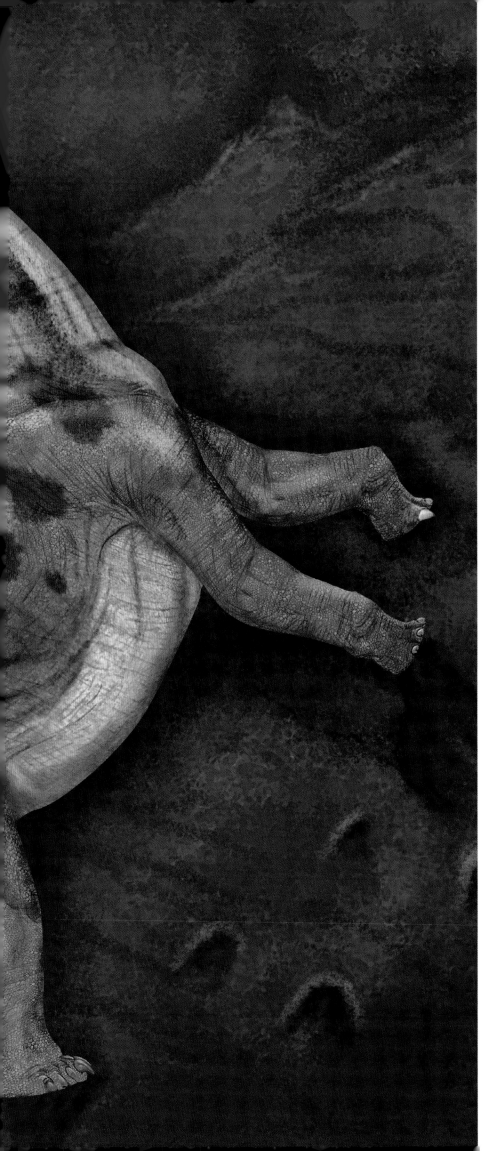

75 Million Years Ago
Present-Day Mongolia

As the carnivorous and herbivorous dinosaurs were fiercely competing against each other, plants were also evolving seventy-five million years ago in what is now Mongolia. During this time many new species emerged to cope with the threats of animals that ate them. Many herbivorous dinosaurs were heading towards extinction due to vegetation changes which they were unable to adapt to. For example, this *Opisthocoelicaudia* lying on the ground had inadvertently consumed an unfamiliar and poisonous flowering plant.

This was a different death, not one from combat. This was a sign of more bad things to come. It was a warning that apart from dinosaurs and other animals, deadly dangers loomed ahead. Subsequently, death from poisoning became more and more common. Unfortunately, the dominant dinosaurs treated this as an isolated event and remained blind to these new threats.

75 million years ago
Present-Day Canada

A *Centrosaurus* with a conspicuous sharp horn growing out of its nose moved out of the darkness that hung over present-day Canada. These dinosaurs' horns marked their individuality. Some were bent forward, some backward, and some were S-shaped pointing upward.

The *Ceratopsidae* family diversified rapidly in the Late Cretaceous period. The number of members in the family also increased. These dinosaurs lived in large social groups, and they were about to become the most powerful herbivorous dinosaurs.

183

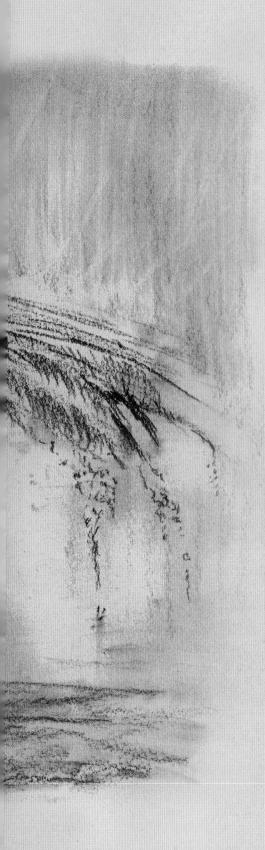

75 Million Years Ago
Present-Day Mongolia

It was raining! The rain came unexpectedly. The skies were clear just a few moments before. Large raindrops began to pour down, soaking a group of young *Oviraptors* playing on the ground. It was the first time that these small dinosaurs had met a heavy rain, and they were anxious, unsure what to do. One of them was petrified and began to cry.

The mother *Oviraptor* arrived, smiled, and stretched her huge forelimbs out to cover the young ones.

The small *Oviraptors* curled up under their mother's protection, and they began to laugh happily.

This heart-warming scene happened seventy-five million years ago in present-day Mongolia.

70 Million Years Ago
Present-Day India

An *Indosuchus* sat basking in the sun in present-day India seventy million years ago.

This was its favorite way to spend its leisure time. It was relaxed, not thinking about anything, not enticed by prey. Most importantly, there were no enemies to disturb its peace of mind.

Indosuchus and *Carnotaurus* belonged to the same family, but compared to the brutal *Carnotaurus*, the six-meter-long *Indosuchus* was much gentler.

70 Million Years Ago
Present-Day Mongolia

It was seventy million years ago, in present-day Mongolia, where a monstrous-looking dinosaur walked in an open field by itself. It had a pair of strong arms, which reached up to 2.4 meters, with long, sharp claws. It was ten meters long, covered in feathers, and had a sail on its back similar to that of *Spinosaurus* or *Ouranosaurus*.

No one was accompanying it. Its solitary figure cast a lonely shadow over the empty, barren plains.

Nevertheless, it did not seem to mind. It could run with incredible speed, and its agile claws were always ready to snatch leaves. This dinosaur was a *Deinocheirus*, the largest member of the *Ornithomimidae* family. Until very recently, the only parts of this animal that were known were their massive forelimbs. New specimens have allowed us to construct a much better picture of what this creature looked like.

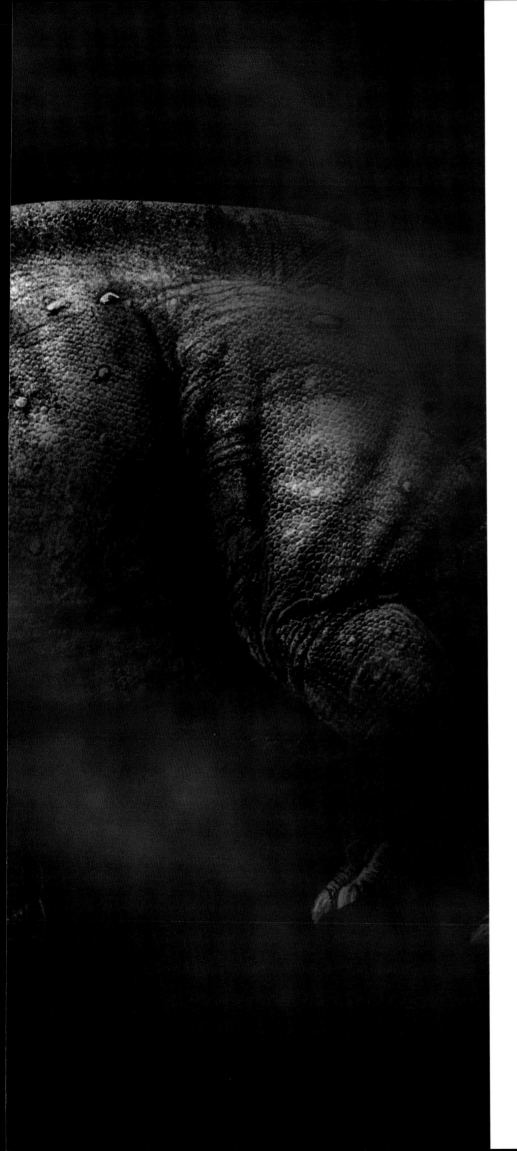

70 Million Years Ago
Present-Day China

Herbivorous dinosaurs always tried to develop new ways to repel carnivorous ones. Over the generations, they changed dramatically.

This was a *Sinoceratops* which lived in Shandong, China, seventy million years ago. It was a strong, powerful herbivorous dinosaur, nearly seven meters long.

However, its size was not the only defense against attacks. What made it feel secure was its array of intimidating horns on its head. Out of its nose, one huge horn grew. Behind its head, there was a large neck frill covered with a layer of bones, with thirteen spikes protruding out along the frill's edges.

The horns and spikes of the *Sinoceratops* were perhaps an effective deterrent against many carnivorous dinosaurs.

The *Ceratopsidae* family originated in present-day Asia. However, the species found in Asia were relatively primitive and quite small in comparison to more advanced and larger specimens found in North America. However, the *Sinoceratops* was an exception, being the only advanced member discovered in Asia.

70 Million Years Ago
Present-Day America

In the world of dinosaurs, beauty had no advantage in the battle-field. A dinosaur's position in the hierarchy was solely determined by its strength.

A *Pachycephalosaurus* had its head and cheeks covered with bony knobs and short bony spikes. It had a bony dome atop its skull, which was twenty-five centimeters thick and tough. There were also bony knobs and short bony spikes around the dome. All these spikes made it look like a dragon in European medieval mythology.

The *Pachycephalosaurus*, of course, would not draw this analogy nor did it care much about it looked. It relied on its keen vision, sensitive hearing, and its courageous spirit to deal with all difficulties. It was one of the most remarkable herbivorous dinosaurs of the Cretaceous period, seventy million years ago, in present-day North America.

70 Million Years Ago
Present-Day Madagascar

The *Abesauridae* family ruled over the southern continents during the Late Cretaceous period, with many members being fierce and tyrannous. This *Majungasaurus*, which lived on what is now the African island of Madagascar, was one of them. When it got really hungry, it would devour anything in its path, even eating its own offspring.

Hunger is the number one enemy all dinosaurs fought against. For generations, dinosaurs had striven to overcome hunger, and cannibalism was common. Still, this could be something difficult for us humans to imagine.

70 Million Years Ago
Present-Day China

In present-day Henan, China, two *Yulong* ran merrily together. Despite the ground being slippery due to the wet lichen, they moved gracefully like ballet dancers. The smaller of the two playfully picked up a fern from the ground. They looked nothing like the majestic dinosaurs, but more like two furry chicks with parrot-like beaks. It was obvious that the *Yulong* was a member of *Oviraptorosauria*. They were the smallest member of the family and by far the cutest!

70 Million Years Ago
Present-Day China

Qiupalong was *Yulong*'s neighbor. It too lived in present-day Henan, China, seventy million years ago.

Qiupalong had gorgeous feathers and a long, beautiful tail. It was busy practicing jumping and running with its head proudly held high.

It supported itself by pressing down the three powerful claws on its hind limbs, bending its hind legs slightly, and then leaping high in the air. As it jumped up, its elegant feathers fluttered.

The *Qiupalong* was an ostrich dinosaur (family *Ornithomimidae*); it is one of the fastest-running dinosaurs, but even so, it still regularly practiced every day, striving to run faster and jump higher.

70 Million Years Ago
Present-Day China

It was a cool and refreshing morning. Although it was still early, an *Xixiasaurus* was already awake, ready to go foraging. The leaves were delicately covered in the morning dew. One light touch would send them dropping down. Even if the *Xixiasaurus* carefully avoided the dewy leaves, the moist air still made its feathers wet and sticky. Nevertheless, that was a small price to pay to get food as early as possible!

The *Xixiasaurus* used its large eyes to survey its surroundings, making sure not to miss any signs of food.

It was a member of the *Troodontidae* dinosaur family, and it lived seventy million years ago in present-day Henan, China.

70 Million Years Ago
Present-Day China

Dinosaurs were born 234 million years ago, but their diversity peaked in the Late Cretaceous period. Many types of herbivorous and carnivorous dinosaurs emerged after millions of years of evolution.

It was seventy million years ago in present-day Shandong, China.

The gigantic *Shantungosaurus* could grow to fifteen meters, challenged the size limits of the *Hadrosauridae* family. The ornate *Sinoceratops* were an advanced Neoceratopsian expanding towards present-day Asia. The evolution of carnivorous dinosaurs produced the ferocious *Zhuchengtyrannus*, which were endowed with sharp teeth and claws and could easily kill a huge *Shantungosaurus*.

The dinosaur family has ushered in a new era of prosperity, unaware that their extinction would be dangerously close.

66 Million Years Ago
Present-Day China

The *Zhuchengtyrannus, which* looked very similar to the contemporaneous Tyranno-saurs, lived in present-day North America. However, the *Qianzhousaurus* which lived sixty-six million years ago in present-day Jiangxi, China, had a different look.

Such marked differences showed the diversity in the *Tyrannosauridae* family during the Late Cretaceous period. They were abundant in number, had many types, and were widely distributed.

The *Qianzhousaurus* lived in the jungle. The lush trees provided camoflauge, which allowed them to hunt with ease. Even with-out the dense foliage, a *Qianzhousaurus* would not have easily been noticed by its prey, because unlike *Zhuchengtyrannus*'s short, wide face, its head was slender, allowing it to hide easily.

This was a common scene: a *Nankangia*, a member of the *Oviraptoridae* family, lost contact with its companions. It was scared and desperate, wanting to find its friends as soon as possible. But it was already in great danger. An well-camouflaged *Qianzhousaurus* appeared behind it. By the time the *Nankangia* was aware of the enemy, it was too late to escape.

66 Million Years Ago
Present-Day China

In present-day Jiangxi, China, three *Nankangia* gathered together to search for food in a big forest to search for food.

"Wow, this forest is really big!"

"Yes, well, there must be plenty of food here!"

These three little ones had traveled a long way from home, and the forest was new to them. They went out together to feel more secure. Luckily, the forest was full of lush vegetation. The three were enticed by the abundant food. "This is good. Oh, that one is also delicious!" The three of them became preoccupied, voraciously stuffing their mouths with all the forest's offerings.

Soon, their stomachs were laden with food. It was only then that they all realized that they had moved away from one another. As the sky turned dark, the *Nankangia* anxiously called out to each other. They were aware that the terrifying *Qianzhousaurus* lived in the forest, and if they were not careful, they might soon become its next meal.

Two *Nankangia* soon found each other. However, as they set out to look for the third, a scream suddenly filled the air. They were frightened and stared intensely into the distance, as they were sure that the bloodcurdling scream came from their friend.

66 Million Years Ago
Present-Day America

As the dinosaurs moved into the last phase of their dominance, both herbivorous and carnivorous ones had became incredibly developed. They had perfected themselves over the ages, and the new species were much better at attacking and defending.

The seven-meter-long *Euoplocephalus*, though not as large as an *Ankylosaurus*, was still a large member of the *Ankylosauridae* family. Its body was covered with bone plates and spikes, including bony eyelids. Its tail had a heavy club. These weapons were not merely for passive defense; it often used them to strike against the most powerful enemies.

These Late Cretaceous Ankylosaurs lived in present-day North America and proved that herbivorous dinosaurs could become fearsome fighters. Moreover, *Ankylosaurus*, the other member of the same family, made fighting even more intense. We will look at its fighting shortly.

66 Million Years Ago
Present-Day America

As herbivorous dinosaurs developed, this naturally led to the evolution of carnivorous ones, with the appearance of the ultimate tyrant of the Mesozoic era, the *Tyrannosaurus rex*.

Tyrannosaurus rex was the most powerful land animal ever. Its sharp teeth could easily pierce prey's scales and flesh and crush the bones. Its terrifying claws could instantly rip open enemies' stomachs. And its large and stout body proved handy in defeating opponents.

It was truly the ultimate predator in the Mesozoic Era. It would never back down from any other dinosaurs!

66 Million Years Ago
Present-Day America

Though powerful, Ankylosaurs were not the ultimate forms in herbivorous dinosaurs' evolution. That honor must go to the *Triceratops*, who had the best weapons.

Triceratops no longer relied on plates for defense. It had horns on its face, including one-meter-long brow horns and a shorter but powerful nasal horn. These horns could scare away its enemies. In addition, its head frill and and the spikes on its back protected all parts of its body.

The *Triceratops* feared no predators, and they had come up with a great dream, which its ancestors could have never considered.

66 Million Years Ago
Present-Day America

A protracted war waged between these three dominant members of the dinosaur world: *Ankylosaurus*, *Triceratops*, and *Tyrannosaurus rex*. The herbivorous dinosaurs could fight on almost equal footing.

Sixty-six million years ago in present-day America, a drying river had spurred a large number of herbivorous dinosaurs to migrate north in search of food. A *Tyrannosaurus rex* came out of the distant forest and caught sight of a stray adult *Triceratops*. This was a great opportunity to attack! The *Tyrannosaurus rex* picked up its pace; with its strong forelimbs and sharp teeth, it easily subdued the enormous prey.

It was not much of a battle for the *Tyrannosaurus rex*. It was too easy and rather unexciting. But the outcome of the battle excited the *Quetzalcoatlus* hovering over its head and the *Bambiraptor* running around at its fee. They waited in eager anticipation for some leftover meat after the tyrant took its share.

The *Tyrannosaurus rex* lowered its head to take a bite from its prey. It focused on enjoying its meal, not bothered by the smaller minions waiting nearby. The leftovers from a *Tyrannosaurus rex*'s meal would feed many smaller animals.

66 Million Years Ago
Present-Day America

As previously mentioned, the balance of power in a war could often be reversed. An *Ankylosaurus*, the largest Ankylosaur, was attacked by a *Tyrannosaurus rex*.

The fighting had been going on for some time, and the *Ankylosaurus* just managed to escape a deadly blow from the attacker. The battle appeared to be clearly favoring the *Tyrannosaurus rex*. Emboldened by the prospect of winning, the *Tyrannosaurus rex* moved its position.

The *Tyrannosaurus rex* did not realize that it had committed a grievous mistake by trying to flank the *Ankylosaurus*, as it now exposed its body to the tail of the *Ankylosaurus*. The *Ankylosaurus* instinctively swung its tail club, and the fifty-kilogram club smashed the leg of the *Tyrannosaurus rex*. The *Tyrannosaurus rex* wanted to retaliate, but its foot was injured, and it fell to the ground.

An injured leg meant that the *Tyrannosaurus rex* could no longer fight, and the *Ankylosaurus* had disappeared.

The *Tyrannosaurus rex*, lying on the ground, could not figure out how it lost to an herbivore.

66 Million Years Ago

Everything seemed to be business as usual. The carnivorous dinosaurs were led by *Tyrannosaurus rex*, and the herbivorous dinosaurs were led by *Triceratops* and *Ankylosaurus*. They were fighting each other, hoping to gain complete dominance and evolve further to be stronger, faster, and quicker. They probably were not aware of the changing geological conditions or how that would affect them.

But the world was changing quickly. Giant volcanos, far bigger than modern ones, were erupting in what is now India. These injected lots of carbon dioxide into the atmosphere. Large inland seas in Asia and North America were drying up, causing winters to be colder and summers hotter. Yet, a far greater disaster was approaching.

A meteorite was about to hit Earth.

Other kinds of animals lived with dinosaurs in this period: modern kinds of lizards, turtles, and crocodiles were present, as well as frogs, salamanders, mammals, and many familiar kinds of plants (like magnolias, pines, and cypress trees) populated forests and swamps. Birds (living dinosaurs) lived at the same time as dinosaurs. All these would eventually survive the catastrophe.

66 Million Years Ago

Dinosaurs used to disdain other reptiles. But their pride was about to come to an end as the disaster loomed. The dinosaurs were under the illusion that bad events were rare and remained convinced that they could maintain their hegemony in the same way as they emerged victorious from battles. However, this time they were wrong. They ruled the world for nearly 170 million years and were the strongest animals to live on Earth, but still they were incapable of defeating nature. Soon their glory would fade, and they would vanish, only leaving fossils as a trace of their existence.

148. Turner, Alan H., D. Pol, J. A. Clarke, G. M. Erickson, and M. Norell. 2007. "A basal dromaeosaurid and size evolution preceding avian flight." *Science* 317 (5843): 1378–81.

149. Turner, A. H., P. J. Makovicky, and M. A. Norell. 2007. "Feather quill knobs in the dinosaur *Velociraptor*." *Science* 317 (5845): 1721.

150. Xu, X., K. Wang, K. Zhang, Q. Ma, L. Xing, C. Sullivan, D. Hu, S. Cheng, S. Wang, et al. 2012. "A gigantic feathered dinosaur from the Lower Cretaceous of China." *Nature* 484 (7392): 92–95.

151. Xu, X., Z.-L. Tang, and X.-L. Wang. 1999. "A therizinosauroid dinosaur with integumentary structures from China." *Nature* 399 (6734): 350–54.

152. Xu, X., Y. Cheng, X.-L. Wang, and C. Chang. 2003. "Pygostyle-like structure from *Beipiaosaurus* (Theropoda, Therizinosauroidea) from the Lower Cretaceous Yixian Formation of Liaoning, China." *Acta Geologica Sinica* 77 (3): 294–98.

153. Zhou, Z. 2006. "Evolutionary radiation of the Jehol Biota: chronological and ecological perspectives." *Geological Journal* 41: 377–93.

154. Xu, X., Y.-N. Cheng, X.-L. Wang, and C.-H. Chang. 2002. "An unusual oviraptorosaurian dinosaur from China." *Nature* 419: 291–93.

155. Xu, X., X.-L. Wang, and H.-L. You. 2000. "A primitive ornithopod from the Early Cretaceous Yixian Formation of Liaoning." *Vertebrata PalAsiatica* 38 (4): 318–25.

156. Zheng, X.-T., H.-L. You, X. Xu, and Z.-M. Dong. 2009. "An Early Cretaceous heterodontosaurid dinosaur with filamentous integumentary structures." *Nature* 458 (19): 333–36.

157. Han, Feng-Lu, Paul M. Barrett, Richard J. Butler, and Xing Xu. 2012. "Postcranial anatomy of Jeholosaurus shangyuanensis (Dinosauria, Ornithischia) from the Lower Cretaceous Yixian Formation of China." *Journal of Vertebrate Paleontology* 32 (6): 1370–95.

158. Han, Gang, Luis M. Chiappe, Shu-An Ji, Michael Habib, Alan H. Turner, Anusuya Chinsamy, Xueling Liu, and Lizhuo Han. 2014. "A new raptorial dinosaur with exceptionally long feathering provides insights into dromaeosaurid flight performance." *Nature Communications* 5: 4382.

159. Ji, Q., P. J. Currie, M. A. Norell, and S. Ji. 1998. "Two feathered dinosaurs from northeastern China." *Nature* 393 (6687): 753–61.

160. Zhou, Z., X. Wang, F. Zhang, and X. Xu. 2000. "Important features of *Caudipteryx* - Evidence from two nearly complete new specimens." *Vertebrata Palasiatica* 38 (4): 241–54.

161. Ji, S., Q. Ji, J. Lu, and C. Yuan. 2007. "A new giant compsognathid dinosaur with long filamentous integuments from Lower Cretaceous of Northeastern China." *Acta Geologica Sinica*, 81 (1): 8–15.

162. Wang, X.-L., and X. Xu. 2001. "A new iguanodontid (*Jinzhousaurus yangi* gen. et sp. nov.) from the Yixian Formation of western Liaoning, China." *Chinese Science Bulletin* 46: 1669–72.

163. Swisher, Carl C., Yuan-qing Wang, Xiao-lin Wang, Xing Xu, and Yuan Wang. 1999. "Cretaceous age for the feathered dinosaurs of Liaoning, China." *Nature* 400: 58–61.

164. Xu, X., Z. Zhou, and R. O. Prum. 2001. "Branched integumental structures in *Sinornithosaurus* and the origin of feathers." *Nature* 410 (6825): 200–4.

165. Hwang, S. H., M. A. Norell, Q. Ji, and K. Gao. 2004. "A large compsognathid from the Early Cretaceous Yixian Formation of China." *Journal of Systematic Palaeontology* 2 (1): 13–30.

166. Xu, X., and M. A. Norell. 2006. "Non-Avian dinosaur fossils from the Lower Cretaceous Jehol Group of western Liaoning, China." *Geological Journal* 41: 419–37.

167. Xu, X., M. A. Norell, W. Xiao-lin, P. J. Makovicky, and W. Xiao-chun. 2002. "A basal troodontid from the Early Cretaceous of China." *Nature* 415: 780–84.

168. Lü, Junchang, Li Xu, Yongqing Liu, Xingliao Zhang, Songhai Jia, and Qiang Ji. 2010. "A new troodontid (Theropoda: *Troodontidae*) from the Late Cretaceous of central China, and the radiation of Asian troodontids." *Acta Palaeontologica Polonica* 55 (3): 381–88.

169. Lü, J., L. Xu, X. Jiang, S. Jia, M. Li, C. Yuan, X. Zhang, and Q. Ji. 2009. "A preliminary report on the new dinosaurian fauna from the Cretaceous of the Ruyang Basin, Henan Province of central China." *Journal of the Palaeontological Society of Korea* 25: 43–56.

170. Buffetaut, E., V. Suteethorn, and H. Tong. 1996. "The earliest known tyrannosaur from the Lower Cretaceous of Thailand." *Nature* 381: 689–91.

171. Lü, J., L. Xu, H. Pu, X. Zhang, Y. Zhang, S. Jia, H. Chang, J. Zhang, and X. Wei. 2013. "A new sauropod dinosaur (Dinosauria, Sauropoda) from the late Early Cretaceous of the Ruyang Basin (central China)." *Cretaceous Research* 44: 202.

172. Xu, X., X.-L. Wang, and H.-L. You. 2001. "A juvenile ankylosaur from China." *Naturwissenschaften* 88 (7): 297–300.

173. Lü, J., L. Xu, S. Jia, X. Zhang, J. Zhang, L. Yang, H.-L. You, and Q. Ji. 2009. "A new gigantic sauropod dinosaur from the Cretaceous of Ruyang, Henan, China." *Geological Bulletin of China* 28 (1): 1–10.

174. Brusatte, S. L., M. A. Norell, T. D. Carr, G. M. Erickson, J. R. Hutchinson, A. M. Balanoff, G. S. Bever, J. N. Choiniere, et al. 2010. "Tyrannosaur paleobiology: new research on ancient exemplar organisms." *Science* 329 (5998): 1481–85.

175. Ji, Q., S.-A. Ji, and L.-J. Zhang. 2009. "First large tyrannosauroid theropod from the Early Cretaceous Jehol Biota in northeastern China." *Geological Bulletin of China* 28 (10): 1369–74.

176. Allain, R., T. Xaisanavong, P. Richir, and B. Khentavong. 2012. "The first definitive Asian spinosaurid (Dinosauria: Theropoda) from the early cretaceous of Laos." *Naturwissenschaften* 99 (5): 369–77.

177. Senter, Phil. 2006. "Comparison of Forelimb Function Between *Deinonychus* and *Bambiraptor* (Theropoda: *Dromaeosauridae*)." *Journal of Vertebrate Paleontology* 26 (4): 897–906.

178. Benton, M. J., S. Bouaziz, E. Buffetaut, D. Martill, M. Ouaja, M. Soussi, and C. Trueman. 2000. "Dinosaurs and other fossil vertebrates from fluvial deposits in the Lower Cretaceous of southern Tunisia." *Palaeogeography, Palaeoclimatology, Palaeoecology* 157 (3–4): 227–46.

179. Dong, Z.-M. 2002. "A new armored dinosaur (Ankylosauria) from Beipiao Basin, Liaoning Province, northeastern China." *Vertebrata PalAsiatica* 40 (4): 276–85.

180. You, H.-L., Q. Ji, J. Li, and Y. Li. 2003. "A new hadrosauroid dinosaur from the mid-Cretaceous of Liaoning, China." *Acta Geologica Sinica* 77 (2): 148–54.

181. Coria, R. A., and P. J. Currie. 2006. "A new carcharodontosaurid (Dinosauria, Theropoda) from the Upper Cretaceous of Argentina." *Geodiversitas* 28 (1): 71–118.

182. Benson, R. B. J., M. T. Carrano, and S. L. Brusatte. 2010. "A new clade of archaic large-bodied predatory dinosaurs (Theropoda: *Allosauroidea*) that survived to the latest Mesozoic." *Naturwissenschaften* 97 (1): 71–78.

183. Benson, R. B., and X. Xu. 2008. "The anatomy and systematic position of the theropod dinosaur *Chilantaisaurus tashuikouensis* Hu, 1964 from the Early Cretaceous of Alanshan, People's Republic of China." *Geological Magazine* 6.

184. Lü, Junchang, Yoichi Azuma, Chen Rongjun, Zheng Wenjie, and Jin Xingsheng. 2008. "A new titanosauriform sauropod from the early Late Cretaceous of Dongyang, Zhejiang Province." *Acta Geologica Sinica* (English Edition) 82 (2): 225–35.

185. Mannion, P. D., P. Upchurch, R. N. Barnes, and O. Mateus. 2013. "Osteology of the Late Jurassic Portuguese sauropod dinosaur *Lusotitan atalaiensis* (Macronaria) and the evolutionary history of basal titanosauriforms." *Zoological Journal of the Linnean Society* 168: 98–206.

186. Kurzanov, Sergei M. 1976. "A new Late Cretaceous carnosaur from Nogon–Tsav, Mongolia." *The Joint Soviet-Mongolian Paleontological Expedition Transactions* (in Russian) 3: 93–104.

187. Russell, D. A. 1972. "Ostrich dinosaurs from the Late Cretaceous of Western Canada." *Canadian Journal of Earth Sciences* 9: 375–402.

188. Young, C.-C. 1958. "The dinosaurian remains of Laiyang, Shantung." *Palaeontologia Sinica*, New Series C 42 (16): 1–138.

189. Xu, Xing, J. Choinere, M. Pittman, Q. Tan, D. Xiao, Z. Li, L. Tan, J. Clark, M. Norell, D. W. E. Hone, and C. Sullivan. 2010. "A new dromaeosaurid (Dinosauria: Theropoda) from the Upper Cretaceous Wulansuhai Formation of Inner Mongolia, China." *Zootaxa* (2403): 1–9.

190. Lü, J., P. J. Currie, L. Xu, X. Zhang, H. Pu, and S. Jia. 2013. "Chicken-sized oviraptorid dinosaurs from central China and their ontogenetic implications." *Naturwissenschaften* 100 (2): 165–75.

Index

Dinosaurs

Biotas

ZHAO Chuang and YANG Yang

&

PNSO's Scientific Art Projects Plan: Stories on Earth (2010–2070)

ZHAO Chuang and YANG Yang are two professionals who work together to create scientific art. Mr. ZHAO Chuang, a scientific artist, and Ms. YANG Yang, an author of scientific children's books, started working together when they jointly founded PNSO, an organization devoted to the research and creation of scientific art in Beijing on June 1, 2010. A few months later, they launched Scientific Art Projects Plan: Stories on Earth (2010–2070). The plan uses scientific art to create a captivating, historically accurate narrative. These narratives are based on the latest scientific research, focusing on the complex relationships between species, natural environments, communities, and cultures. The narratives consider the perspectives of human civilizations while exploring Earth's past, present, and future. The PNSO founders plan to spend sixty years to do research and create unique and engaging scientific art and literature for people around the world. They hope to share scientific knowledge through publications, exhibitions, and courses. PNSO's overarching goal is to serve research institutions and the general public, especially young people.

PNSO has independently completed or participated in numerous creative and research projects. The organization's work has been shared with and loved by thousands of people around the world. PNSO collaborates with professional scientists and has been invited to many key laboratories around the world to create scientific works of art. Many works produced by PNSO staff members have been published in leading journals, including *Nature*, *Science*, and *Cell*. The organization has always been committed to supporting state-of-the-art scientific explorations. In addition, a large number of illustrations completed by PNSO staff members have been published and cited in hundreds of well-known media outlets,

including the *New York Times*, the *Washington Post*, the *Guardian*, *Asahi Shimbun*, the *People's Daily*, BBC, CNN, Fox News, and CCTV. The works created by PNSO staff members have been used to help the public better understand the latest scientific discoveries and developments. In the public education sector, PNSO has held joint exhibitions with scientific organizations including the American Museum of Natural History and the Chinese Academy of Sciences. PNSO has also completed international cooperation projects with the World Young Earth Scientist Congress and the Earth Science Matters Foundation, thus helping young people in different parts of the world understand and appreciate scientific art.

KEY PROJECTS

I. Darwin: An Art Project of Life Sciences
*The models are all life-sized and are based on fossils found around the world
1.1 Dinosaur fossils
1.2 Pterosaurs fossils
1.3 Aquatic reptile fossils
1.4 Ancient mammals of the Cenozoic Era
1.5 Chengjiang biota: animals in the Early Cambrian from fossils discovered in Chengjiang, Yunnan, China
1.6 Jehol biota: animals in the Mesozoic Era from fossils discovered in Jehol, Western Liaoning, China
1.7 Early and extinct humans
1.8 Ancient animals that coexisted with early and extinct humans
1.9 Modern humans
1.10 Animals of the *Felidae* family
1.11 Animals of the *Canidae* family
1.12 Animals of the Proboscidea order